# Precalculus
## and
# Computer Programming

## Other Books by Francis D. Hauser

*Excel with VBA for Engineers and Mathematicians*, 2015. This book includes programs for computing eigenvalues, transfer functions, frequency response, root locus, and Dantzig's simplex algorithm.

*The Golden Ratio: The Facts and the Myths*, 2015. This book looks at the golden ratio from an engineer's viewpoint.

*Eigenvalues and Eigenvectors*, 2016. This book includes complete listings of two programs: one that computes real and complex eigenvalues; and one that computes the corresponding eigenvectors.

# Precalculus
## and
# Computer Programming

USING

# Excel with VBA

(PC and Mac)

# Francis D. Hauser

*Precalculus and Computer Programming Using Excel with VBA*

Copyright © 2017 Francis D. Hauser. All rights reserved.

Excel ® is a registered trademark of Microsoft Corporation.

ISBN-13: 978-1548073640
ISBN-10: 1548073644

Library of Congress Control Number: 2017909468

CreateSpace Independent Publishing Platform, North Charleston, South Carolina.

# Abstract

The reader can expect to be ready for calculus and to write computer programs to help learn math. All topics of precalculus are covered. The programs that do the math in each chapter are listed and studied.

The exponential, logarithmic, and trigonometric functions are defined. Because they are not algebraic, electronic calculators must use approximation methods. The reader is introduced to the use of an infinite series to compute them. For several functions, programs are written to compare results between a finite Taylor series and a calculator.

One discussion in the book is about the base e exponential function. It's the only exponential function whose slope is equal to the function itself. This leads to a calculus formula that applies when the rate of change of a variable is directly proportional to the response of the variable. This calculus formula is applied to a spring-mass-damper system. This leads to the derivation of a polynomial, and shows how zeros of this polynomial affect response of the mass.

Another discussion shows how computing the determinant of a matrix yields the zeros of a polynomial. This discussion leads to a computer program that finds all the zeros of a general polynomial. This program is listed, explained, and demonstrated. The program can find all the zeros of a fourteenth-order Butterworth polynomial.

The computer programs use Excel with VBA, which is inside the Microsoft Office program. Each chapter concludes with listings and discussions of the VBA language used to write them. The reader can execute these and modify them to study future problems.

## Acknowledgments

To Melco Absin and Angie Ilagan-Absin__thank you both sincerely.
To Allan and Aldiv Divinagracia__thank you both sincerely.
To Mary and James Gustafson__thank you both sincerely.

# Contents

Appendices

# Introduction and Summary

My goal while writing this book was to prepare the reader for calculus. Since the material is ideally suited, my other goal was to use it to show the reader how to write computer programs. Each chapter concludes with a discussion of those parts of the VBA language that are needed to write the programs that do the math in that chapter. Complete listings of the programs are included. To learn math together with computer programming is a win-win situation. There is hardly a better way to learn math than to copy and modify programs that solve useful math problems.

The programs use Excel with VBA that is included in the Microsoft Office software. For a mathematician and engineer, this computer language is equivalent to C++, FORTRAN or MATLAB. It's a twenty-first-century version of BASIC.

The following is a summary of the book.

Chapter 1 is on complex numbers. The discussion emphasizes that $\sqrt{-1}$ is a legitimate number. A polynomial is plotted in three dimensions in order to see its complex zeros.

Chapter 2 is on exponential and logarithmic functions. These are inverses of each other. As part of calculus preparation, each function is computed from its Taylor series approximation. Results are compared with calculator results, which are themselves approximations. The base e exponential function is called the *natural* function. It's the only exponential function whose slope is equal to the function itself. This yields the calculus formula that applies when the rate of change of a variable is directly proportional to the response of the variable itself.

Chapter 3 is on linear equations and matrices. The topics include matrix inversion and several ways to solve linear equations (method of substitution, row operations, and Cramer's rule). Determinants and homogeneous equations are discussed.

Chapter 4 is on polynomials and the meaning of zeros. The calculus formula discussed in chapter 2 is applied to a spring-mass-damper system. This leads to the derivation of a polynomial, and shows how zeros of this polynomial affect the response of the mass. Depending on where the zeros are located, the response is convergent or divergent, and oscillatory or damped. This chapter concludes with Descartes' rule of signs and the Routh array. These show the type and general location of polynomial zeros without actually computing them.

Chapter 5 is on finding all the zeros of a polynomial. It shows how computing the determinant of a matrix yields the zeros of a polynomial. This chapter also shows how to change a polynomial into a matrix. The computer program for finding all the zeros (real and complex) is in appendix B, but chapter 5 shows examples using it.

Chapter 6 is on trigonometry. In the discussion on the trig functions, it's emphasized that even calculators and computers can only approximate the values of the trig functions. The reader is introduced to using infinite series approximations to compute these functions. Taylor series approximations are compared to calculator results.

Chapter 7 is on conics. Equations in polar and rectangular form are derived for centered, translated, and rotated parabolas, ellipses, hyperbolas, and circles.

Chapter 8 is on the VBA computer language. It contains a summary of all the syntax used in the programs in the book. It also contains all of the syntax that was not needed in this book. The topics covered in this chapter include the following:

- The programming environment between Excel and VBA
- Making graphs
- Arrays and data types
- Dynamic arrays and Excel array functions
- Functions
- Looping and branching, and the operators
- The Call statement
- The GoSub statement
- The GoTo statement
- Debugging
- Summary list of the VBA elements

Appendix A: Derivation of the Quadratic formula.

Appendix B: Detailed outline of the QR algorithm for computing polynomial zeros. This includes a demonstration of how it works. Also included is the listing of the program that computes all the zeros (real and complex) of a polynomial.

Appendix C: Listing of the LR program.

Appendix D: Listing of the QR program.

Appendix E: Derivation of the coordinate transformation used to rotate the general conic equation.

# Chapter 1: Complex Numbers

This chapter is about $\sqrt{-1}$. This was not considered a legitimate number until Euler and Gauss wrote their many books around the year 1800.

Let's begin with the second-order polynomial $F = x^2 + bx + c$.

Two values of x make F = 0. These are given by the famous Quadratic formula which is derived in appendix A.

$$x_1 = \frac{-b + \sqrt{b^2 - 4c}}{2} \quad \text{and} \quad x_2 = \frac{-b - \sqrt{b^2 - 4c}}{2}$$

Let's plug in some numbers.

• Case 1 with $b = -4$ and $c = 3$: $\qquad F = x^2 - 4x + 3$

$$x_1 = \frac{4 + \sqrt{16 - 12}}{2} = 3 \quad \text{and} \quad x_2 = \frac{4 - \sqrt{16 - 12}}{2} = 1$$

$x_1$ and $x_2$ are called real zeros of F.

F can be written in factored form. $\qquad F = (x - x_1)(x - x_2)$

From this form, it is directly seen that F = 0 when $x = x_1$ or when $x = x_2$.

The following table shows a plot of F.

| X | F | |
|---|---|---|
| Large negative value | Large positive value | Case 1 |
| Large positive value | Large positive value | |
| 0 | 3 | |
| 1 | 0 | |
| 3 | 0 | |

The plot agrees with the Quadratic formula.

• Case 2 with $b = -4$ and $c = 4$: $\qquad$ $F = x^2 - 4x + 4$

$$x_1 = \frac{4 + \sqrt{16 - 16}}{2} = 2 \quad \text{and} \quad x_2 = \frac{4 - \sqrt{16 - 16}}{2} = 2$$

Again, $x_1$ and $x_2$ are called real zeros of F.

The following table shows a plot of F.

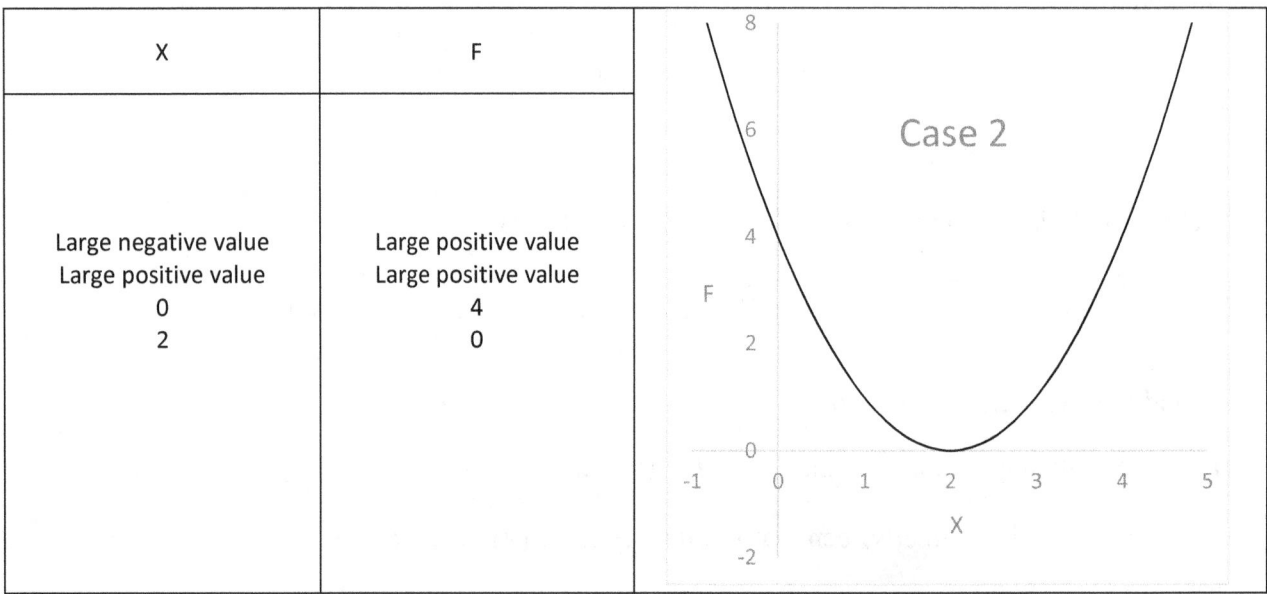

| X | F | |
|---|---|---|
| Large negative value | Large positive value | |
| Large positive value | Large positive value | |
| 0 | 4 | |
| 2 | 0 | |

Again, the plot agrees with the quadratic formula. Because this case requires such precise values on the coefficients of F, it really only occurs in textbooks. In reality, coefficients always have tolerances.

• Case 3 with $b = -4$ and $c = 5$: $\qquad$ $F = x^2 - 4x + 5$

$$x_1 = \frac{4 + \sqrt{16 - 20}}{2} = \frac{4 + \sqrt{-4}}{2}$$

We know that $\sqrt{-4} = \sqrt{-1 * 4} = 2\sqrt{-1}$. Therefore, $x_1 = \frac{4 + 2\sqrt{-1}}{2} = 2 + \sqrt{-1}$. Similarly, $x_2 = 2 - \sqrt{-1}$.

Let's check $x_1$. $F = \left(2 + \sqrt{-1}\right)^2 - 4\left(2 + \sqrt{-1}\right) + 5$.

• The first term is $\left(2 + \sqrt{-1}\right) * \left(2 + \sqrt{-1}\right) = 4 + 2\sqrt{-1} + 2\sqrt{-1} - 1 = 3 + 4\sqrt{-1}$.

• The second term is $-4\left(2 + \sqrt{-1}\right) = -8 - 4\sqrt{-1}$.

• The third term is 5.

• Adding these we get $F = 0 + 0\sqrt{-1}$. This checks out.

Gauss and Euler are both credited with formalizing the rules of algebra. They were among the first to say that $\sqrt{-1}$ is a true number. It is so important that it has been given a special symbol, $\sqrt{-1} = j$. Some mathematicians use $\sqrt{-1} = i$. Note that $j * j = -1$.

Let's check the value of F using the notation $x_2 = 2 - j$.   $F = (2 - j)^2 - 4(2 - j) + 5$.

• The first term is $(2 - j)*(2 - j) = 4 - 2j - 2j - 1 = 3 - 4j$.

• The second term is $-4(2 - j) = -8 + 4j$.

• The third term is 5.

• Adding these we get $F = 0 + 0j$. This checks out.

For Case 3, the factored form is $F = (x - 2 - j)(x - 2 + j) = x^2 - 2x + \cancel{jx} - 2x + 4 - \cancel{j2} - \cancel{jx} + \cancel{j2} + 1$.

This equals $F = x^2 - 4x + 5$. The $j$ terms have cancelled because the coefficients of F are real numbers. We will discuss this later.

For Case 3, $x_1$ and $x_2$ are called complex zeros of F. So let's plot F using complex values for x, namely, $x = \alpha + j\beta$.

$$F = (\alpha + j\beta)^2 - 4(\alpha + j\beta) + 5$$

• The first term is $(\alpha + j\beta)*(\alpha + j\beta) = \alpha^2 + j\alpha\beta + j\alpha\beta + j\beta j\beta = \alpha^2 - \beta^2 + j2\alpha\beta$.

• The second term is $-4(\alpha + j\beta) = -4\alpha - j4\beta$.

• The third term is 5.

• Adding and grouping these, we get $F = (\alpha^2 - \beta^2 - 4\alpha + 5) + j(2\alpha\beta - 4\beta) = F_{real} + jF_{imag}$.

This is a complex number. We need two axes to plot it. The term multiplied by $j$ is called the imaginary part. The other part is called the real part. Because they are independent, these axes are perpendicular.

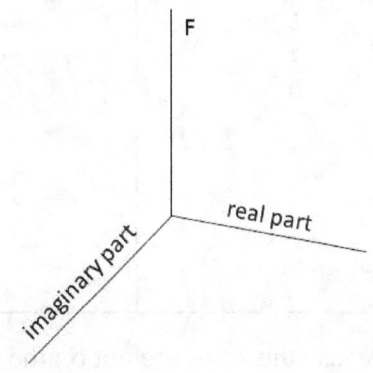

Let's plot the magnitude of F at several values of α and β. From the Pythagorean theorem, $F_{mag} = \sqrt{F_{real}^2 + F_{imag}^2}$ , where $F_{real} = \alpha^2 - \beta^2 - 4\alpha + 5$ and $F_{imag} = 2\alpha\beta - 4\beta$ .

Here's a sample calculation. For α = 4 and β = 2, $F_{mag} = \sqrt{(16-4-16+5)^2 + (16-8)^2} = 8.06$ .

| beta | alpha | | |
|---|---|---|---|
| | 0 | 2 | 4 |
| 0 | 5 | 1 | 5 |
| 1 | 5.66 | 0 | 5.66 |
| 2 | 8.06 | 3 | 8.06 |

From the plot, we see $F_{mag} = 0$ at α = 2 and β = 1. This is one of the complex zeros. To see the zero at $2 - j$, the range for β can be extended to −2.

What happens if we plot $F_{mag}$ using only real values of x? The following plot shows this.

| X | F |
|---|---|
| Large negative value | Large positive value |
| Large positive value | Large positive value |
| 0 | 5 |
| 2 | 1 |
| 4 | 5 |

There is no real axis crossing because the zeros are not on the real axis. They are in the complex plane. But the plot does indicate that a complex zero is present. F bends toward zero around x = 2.

Let's see what happens if we plot Case 1 with x = α + jβ.

$$F = \left(\alpha + j\beta\right)^2 - 4\left(\alpha + j\beta\right) + 3$$

For this case, $F_{mag} = \sqrt{F_{real}^2 + F_{imag}^2}$ , where $F_{real} = \alpha^2 - \beta^2 - 4\alpha + 3$ and $F_{imag} = 2\alpha\beta - 4\beta$ . The following table shows a plot of this $F_{mag}$.

| beta | alpha | | | | |
|---|---|---|---|---|---|
| | 0 | 1 | 2 | 3 | 4 |
| 0 | 3 | 0 | 1 | 0 | 3 |
| 1 | 4.47 | 2.24 | 2 | 2.24 | 4.47 |
| 2 | 8.06 | 5.66 | 5 | 5.66 | 8.06 |

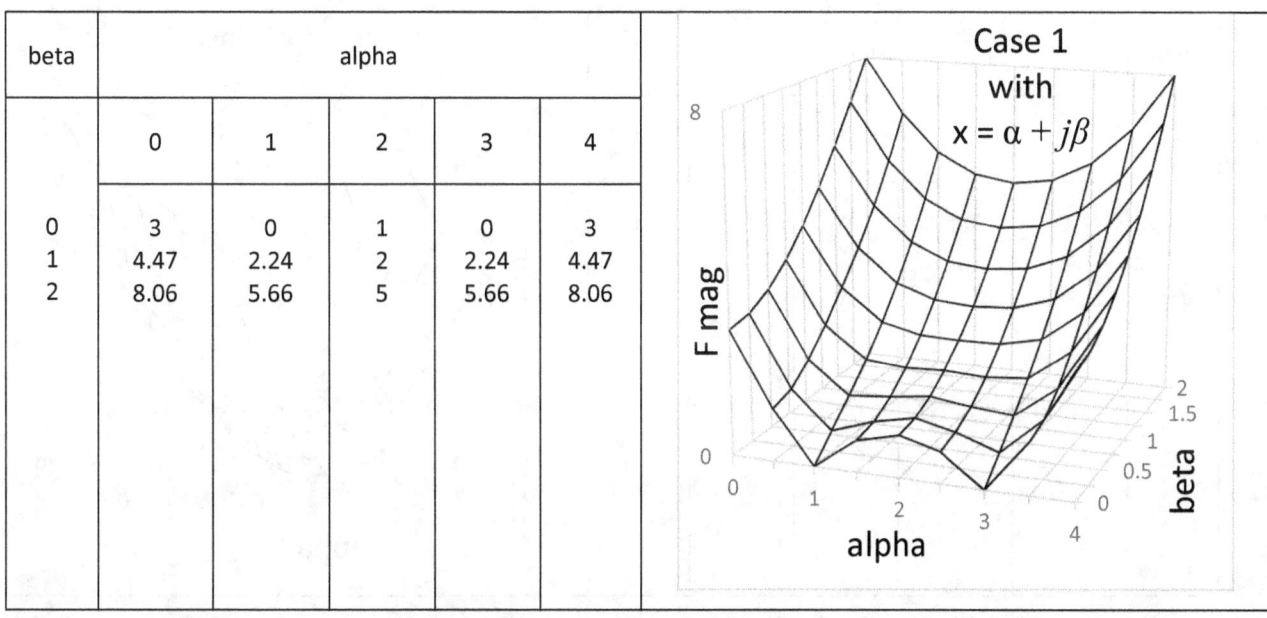

The zeros are at x = 1 and x = 3. We see these zeros on the **alpha** axis where β = 0.

Let's plot Case 2 with $x = \alpha + j\beta$.

$$F = (\alpha + j\beta)^2 - 4(\alpha + j\beta) + 4$$

For this case, $F_{mag} = \sqrt{F_{real}^2 + F_{imag}^2}$, where $F_{real} = \alpha^2 - \beta^2 - 4\alpha + 4$ and $F_{imag} = 2\alpha\beta - 4\beta$. The following table shows a plot of this $F_{mag}$.

| beta | alpha | | | | |
|---|---|---|---|---|---|
| | 0 | 1 | 2 | 3 | 4 |
| 0 | 4 | 1 | 0 | 1 | 4 |
| 1 | 5 | 2 | 1 | 2 | 5 |
| 2 | 8 | 5 | 4 | 5 | 8 |

The zeros for this case are both at $x = 2$. The plot shows zeros at $x = 2 + j0$ and $x = 2 - j0$.

## Complex Zeros Come in Conjugate Pairs

Suppose a second-order polynomial has the following complex zeros:

$$x_1 = a + jb \text{ and } x_2 = c + jd$$

In factored form, this is $F = (x - a - jb)(x - c - jd)$.

In polynomial form, this is $F = x^2 - cx - jdx - ax + ac + jad - jbx + jbc - bd$.

Gathering terms, we get $F = (x^2 - cx - ax + ac - bd) + j(-dx + ad - bx + bc)$.

If the coefficients of F are real, then $-dx + ad - bx + bc = -x(d + b) + (ad + bc) = 0$. This means that $d = -b$ and $ad = -bc$. Therefore, $c = a$. We see that the complex zeros must be $x_1 = a + jb$ and $x_2 = a - jb$. They are a conjugate pair.

This is true for all polynomials with real coefficients. This is because they can be factored into products of first-order terms like $(x + a)$ and second-order terms like $(x^2 + bx + c)$. In precalculus, polynomials will all have real coefficients. Therefore, all complex zeros will be in conjugate pairs.

## Complex Arithmetic

$\sqrt{-1}$ is a number. It is treated the same way as $\sqrt{c}$ in the quantity $\left(a + b\sqrt{c}\right)$.

- Addition: $(a + jb) + (c + jd) = (a + c) + j(b + d)$

- Subtraction: $(a + jb) - (c + jd) = (a - c) + j(b - d)$

- Multiplication: $(a + jb) * (c + jd) = (ac - bd) + j(ad + bc)$

- Division: $Q = \dfrac{a + jb}{c + jd}$. Let's multiply the denominator by its conjugate $(c - jd)$.

$$(c + jd)(c - jd) = c^2 - d^2$$

We have changed the denominator to a real number. To keep $Q$ unchanged, we have to multiply the numerator by $(c - jd)$.

$$Q = \frac{(a + jb)}{(c + jd)} \frac{(c - jd)}{(c - jd)} = \frac{(ac + bd) + j(bc - ad)}{c^2 - d^2} = \left(\frac{ac + bd}{c^2 - d^2}\right) + j\left(\frac{bc - ad}{c^2 - d^2}\right)$$

This is how complex numbers are divided.

- Powers and roots: DeMoivre's theorem is used to compute powers and roots of complex numbers. It is derived and used in chapter 5 to check the accuracy of polynomial zeros.

Note: The square root of a negative real number has two solutions.

$$\sqrt{-a} = +j\sqrt{a} \text{ and } \sqrt{-a} = -j\sqrt{a}$$

Let's verify this. $\left(+j\sqrt{a}\right)^2 = j^2\left(\sqrt{a}\right)^2 = -a$. Also $\left(-j\sqrt{a}\right)^2 = (-1)^2 j^2 \left(\sqrt{a}\right)^2 = -a$.

Also note that $\sqrt{a + jb} \neq \sqrt{a} + \sqrt{jb}$.

We will now discuss **VBA** programming in order to do the math in chapter 1.

# Complex Numbers – Programming in Chapter 1

To learn programming, we have to start with the environment. Where do we put the programs? How do we execute them? How do we get the results?

## The Environment

Open Excel and create a new workbook. Then Save As:

| For this discussion, Save the Workbook as Indicated below |
| --- |
| Filename:  First<br><br>Where:  Desktop<br><br>Type:  Excel Macro-Enabled Workbook (.xlsm) |
| When the file named First is reopened, *click* the Enable Macros button if required. |

On a tab near the bottom-left of the screen, the spreadsheet is called Sheet1. The + next to this tab is for adding more spreadsheets. For now, leave it at Sheet1.

The rows of the spreadsheets are given numbers. The columns are given letters. If you want to change the letters to numbers, see the following table.

| How to Switch the Columns from Letters to Numbers |
| --- |
| On a PC with Excel 2013 or 2016: File → Options → Formulas → R1C1 Reference Style.<br><br>On a Mac with Excel 2011: Excel → Preferences → General → R1C1 Reference Style.<br><br>If these instructions don't work on your system, Google<br>      How to switch to R1C1 reference style in Excel [your version]. |

If the Developer tab is not shown, it must be activated. See the following table.

| How to Activate the Developer Tab |
| --- |
| On a PC with Excel 2013 or 2016: File → Options → Customize Ribbon → Developer.<br><br>On a Mac with Excel 2011: Excel → Preferences → Ribbon → Developer.<br><br>If these instructions don't work on your system, Google<br>      How to activate the Developer tab in Excel [your version]. |

The following is a step-by-step discussion on how to write a VBA program. Click the **Developer** tab. On its ribbon, click **Editor** (or **Visual Basic**). Two things happen.

| 1) The *Project* Window Opens. | 2) The Category *Run* Appears in the Menu Bar. |
|---|---|
| Project<br><br>VBA Project (First.xlsm)<br>• Sheet1 (Sheet1)<br>• ThisWorkbook<br><br>Sheet1(Sheet1) and This Workbook are names of two windows that can contain VBA code. They are currently empty. | Selecting Run → Run Macro, opens the Macros window. It lists the names of programs that can be run right now.<br><br>Macros<br><br>*Initially Empty* |

## Writing a Program

1) In the **Project** window, double-click **ThisWorkbook**. The code window will open.

2) At the cursor in the empty window, enter the following five lines of VBA code.

| First.xlsm: ThisWorkbook (Code) | |
|---|---|
| **Code** | **Comments** |
| Sub one()<br>  Cells(1, 1) = 4 * Atn(1)<br>  pi = Cells(1, 1)<br>  Cells(1, 3) = 2 * pi<br>End Sub | •The name of the program is *one*. The empty parentheses are necessary.<br><br>•The *Cells...* statement writes the value of $\pi = 4*\tan^{-1}(1)$ into spreadsheet cell row 1, column 1.<br><br>•The *pi=...* statement reads the value of $\pi$ from cell row 1, column 1.<br><br>•This *Cells...* statement writes the value of $2*\pi$ into cell row 1, column 3. |

3) In the top menu bar, File→Save and close the window.

4) Select **Run→Run Macro**. This opens the **Macros** window. It will show an item called **ThisWorkbook.one**. Click on this item and then select **Run**.

5) On the spreadsheet, see the following:

- The number $\pi$ is written in the cell row 1, column 1.
- The number $2\pi$ is written in row 1, column 3.

6) Prepare for the next run. Clear the spreadsheet by selecting (highlighting) the data and then clicking **Edit→Clear→All**. Selecting is done by holding-down-the-clicker and dragging.

Via the + at the bottom of the spreadsheet, add **Sheet2**. Then click **Editor (Visual Basic)**. This opens the **Project** window. Observe that **Sheet2(Sheet2)** has been added. The windows **Sheet1(Sheet1)** and **Sheet2(Sheet2)** are currently empty.

| Project |
|---|
| <u>VBA Project (First.xlsm)</u> |
| Sheet1(Sheet1) |
| Sheet2(Sheet2) |
| ThisWorkbook |

Select **Run→Run Macro**. In the **Macros** window, click **ThisWorkbook.one** and **Run**. This time, $\pi$ and $2\pi$ are written on **Sheet 2**. This shows that the program named **one**, in **ThisWorkbook**, reads and writes to the open (visible) spreadsheet. For this discussion, clear **Sheet 2**.

### Adding More Code Windows

We will now add another code window. If the **Project** window is not open, click **Editor**. Then **Insert→Module**. The **Project** window now contains four code windows.

| Project |
|---|
| <u>VBA Project (First.xlsm)</u> |
| Module1 |
| Sheet1(Sheet1) |
| Sheet2(Sheet2) |
| ThisWorkbook |

We will now copy the program named **one** from **ThisWorkbook** to **Module 1**.

- Double-click **ThisWorkbook**, **Edit→Select All→Copy**, and close the window.
- Double-click **Module1**, **Edit→Paste**, **File→Save**, and close the window.

Now, select **Run→Run Macro** to open the **Macros** window.

| Macros |
|---|
| one |
| ThisWorkbook.one |

We can now run two programs. We have shown that **ThisWorkbook.one** uses the open (visible) spreadsheet. In the **Macros** window, if we click on **one** and select **Run**, $\pi$ and $2\pi$ will again appear on **Sheet 2**. We see that **Module 1** also uses the open spreadsheet. More modules can be added. Each **Module** $j$ works the same.

## Notes

• By using ThisWorkbook and the Module*j* code windows, a program can run many cases, changing only the spreadsheets between cases.

• Code windows can contain more than one program.

## Printing the Spreadsheet Data

Spreadsheet data can be printed directly from Excel. Data can also be pasted into Word and PowerPoint. Tip: Paste Options→Paste Special→Bitmap (or PDF).

## More about Code Windows and Spreadsheets

The following table contains more notes. The reader may skip this table and then read it after writing a few programs.

---

• Programs read and write to specific cells. Cutting and pasting data from these cells to others can keep the specific cells free for subsequent runs.

• An attractive feature of the spreadsheet is that, after a run, clarifying comments can be manually added.

• To select two or more areas on the spreadsheet:
  • First area, hold-down-clicker and drag;
  • Subsequent areas, hold-down-command-key while holding-down-clicker and dragging, or hold-down-ctrl-key while holding-down-clicker and dragging.

• All code windows can be given usernames. The Properties window is where this is done. To access this window from the menu bar, View→Properties window.

• On the menu bar is the command View→Code. This opens a code window just like double-clicking on its name.

• The Project window contains all of the files that are open. This facilitates things like copying from one file to another.

• It is good practice to close code windows before selecting Run→Run Macro.

---

We are ready to study the programs that did the math in this chapter. A good way to learn how to program is to copy a simple program that does an important task. Let's look at the program that computes the amplitude of the polynomial used in Case 1 of this chapter.

$$F = x^2 - 4x + 3$$

F will be computed using the following series of VBA statements that are called **For-Next** looping. These statements should be read just like you read a book.

$$\text{For } x = -1 \text{ To } 5 \text{ Step } 0.5$$
$$F = x^2 - 4x + 3$$
$$\text{Next } x$$

For the statements shown, F will be computed for each x from −1 to 5 in increments of 0.5.

On each pass through the loop, the value of x and F will be printed to a spreadsheet. The VBA statements for doing this are

$$\text{Cells(row,column)} = x$$
$$\text{Cells(row,column)} = F$$

The following table shows a listing of the program.

| Program case_1 – Listing and Description | |
|---|---|
| ```
Sub case_1()
i = 1
For x = -1 To 5 Step 0.5
    F = x ^ 2 - 4 * x + 3
    Cells(i, 1) = x
    Cells(i, 2) = F
    i = i + 1
Next x
End Sub
``` | • The empty parentheses is required.<br>• i specifies the row number. This statement initializes i.<br><br>• This is the polynomial. Note the VBA syntax for exponentiation and multiplication.<br>• These are the output statements.<br>• The value of i increases each time F is evaluated.<br><br>• This last statement is required. |
| VBA syntax note: The statement i = i + 1 shows that the equal sign "=" means "is replaced by". | |

Type this program into any module, and run it using the instructions given previously. Don't forget the good practice of closing the code window before selecting Run→Run Macro. The following table shows the output and the plot. Instructions for making the plot will follow.
```

| row | column 1 for X | column 2 for F |
|-----|----------------|----------------|
| 1 | -1 | 8 |
| 2 | -0.5 | 5.25 |
| 3 | 0 | 3 |
| 4 | 0.5 | 1.25 |
| 5 | 1 | 0 |
| 6 | 1.5 | -0.75 |
| 7 | 2 | -1 |
| 8 | 2.5 | -0.75 |
| 9 | 3 | 0 |
| 10 | 3.5 | 1.25 |
| 11 | 4 | 3 |
| 12 | 4.5 | 5.25 |
| 13 | 5 | 8 |

Output From Program Named case_1

Excel refers to this chart as a **Scatter** chart. The horizontal axis data must be on the left. The following two tables show how to make this plot. One table is for a PC and the other is for a Mac.

---

### One Way to Make Scatter Charts on a PC with Excel 2013 and 2016

• Select (highlight) the data to be plotted. Select the Insert Tab→ Charts→ Scatter→ Scatter with Smooth Lines.

• Click the chart to bring up a ribbon with tabs for Design and Format.

(1) Design. Under Add Chart Elements:
   (A) Axes. Options include range, increments, and tick marks.

   (B) Chart Title. Can be edited on Home tab. Can be moved by dragging.

   (C) Axis Titles. Options→Size and Properties→Text direction.

   (D) Gridlines. Can be toggled on or off.

(2) Format. Each series may be a line, or markers, or markers with a line.
   (A) In Current Selection area, select a series. Then Format Selection to open the Format Data Series window.
   Under the paint bucket: (a) Marker. • Marker Options.
                                              • Fill Options.
                                              • Border Options.
              (b) Line. Color, weight, and dash. For markers: (b) Line. Line→No Line.
If necessary, under Series Options, select the next series and repeat. Finally, close the window.

   (B) Size. Type in the desired height and width.

---

### One Way to Make Scatter Charts on a Mac with Excel 2011

• Select (highlight) the data to be plotted. Select the Charts Tab→ Scatter→ Smooth Lined Scatter.

• Click the chart to bring up a ribbon with tabs for Chart Layout and Format.

(1) Chart Layout.
   (A) Axes. Options include scale for range, increments, and tick marks.

   (B) Chart Title. Can be edited on Home tab. Can be moved by dragging.

   (C) Axis Titles. Options→Textbox→Text direction.

   (D) Gridlines. Can be toggled on or off.

(2) Format. Each series may be a line, or markers, or markers with a line.
   (A) In Current Selection area, select a series. Then Format Selection to open the Format Data Series window.
                    (a) Marker Style. Options.
                    (b) Marker Fill. Options.
                    (c) Marker Line. Options are for the border of each marker.
                    (d) Line. Color, weight, and dash. For markers: (d) Line. Color→No Line.
   Click OK to close the Format Data Series window. In Current Selection area, select next series and repeat.

   (B) Size. Type in the desired height and width.

As far as programming is concerned, the math to do Case 3 is the same as Case 1. The difference is the creation of a three-dimensional chart. Discussion of this type of chart is delayed until section 8.2.

## Printing a Chart

A chart can be printed directly from Excel. It can also be pasted into Word and PowerPoint. One way to paste a chart into Word is shown in the following table.

> - Click on the chart and copy.
> - Paste into PowerPoint as a graphic object, and resize, reformat, and annotate if desired.
> - Copy from PowerPoint.
> - Paste into Word via Paste Special → PDF or Paste Special → Bitmap.

Another way to paste a chart into Word is

$$\text{Paste} \rightarrow \text{Paste Special} \rightarrow \text{Graphic Object.}$$

This allows you to alter the appearance of the chart right in Word.

# Chapter 2: Exponential and Logarithmic Functions

The equation studied in this chapter is

$$y = b^x \qquad \text{Equation 2.1}$$

where b is a positive constant called the base and x is a variable called the exponent.

Don't confuse this with polynomials wherein the constant is in the exponent. In precalculus, b is positive and x is real. Therefore, y is real.

We will consider both of the problems that **Equation 2.1** presents.

1. Given x, find y. Mathematically this means to raise b to the x power.
2. Given y, find the exponent x. Mathematically this means to compute the logarithm of y.

Let's start with the equation $y = 2^x$. The notation $2^x$ is shorthand for multiplication. The following table illustrates some of this notation.

| Longhand | Shorthand | The Answer |
|:---:|:---:|:---:|
| $y = 2*2*2$ | $y = 2^3$ | $y = 8$ |
| $y = \dfrac{1}{2*2*2}$ | $y = \dfrac{1}{2^3} = 2^{-3}$ | $y = 0.125$ |
| $y = \dfrac{2*2*2}{2*2*2}$ | $y = \dfrac{2^3}{2^3} = 2^{3-3} = 2^0$ | $y = 1$ |

Let's compute a few values for $y = 2^x$, and compare them to what a calculator would get.

| x | $y = 2^x$ |
|:---:|:---:|
| -3 | 0.125 |
| -2 | 0.25 |
| -1 | 0.5 |
| 0 | 1 |
| 1 | 2 |
| 2 | 4 |
| 3 | 8 |

As expected, they match. Now let's do the reverse problem. If we were given y and wanted to compute x, we could just use this same plot in reverse. Instead, let's compute x directly from y. Mathematically this is called "taking the log of y to base 2".

$$Log_2(y) = Log_2(2^x) = x$$

It's very important to know that **$Log_2(y)$ is the exponent of $2^x$**. Let's do a few numbers and compare them to calculator results.

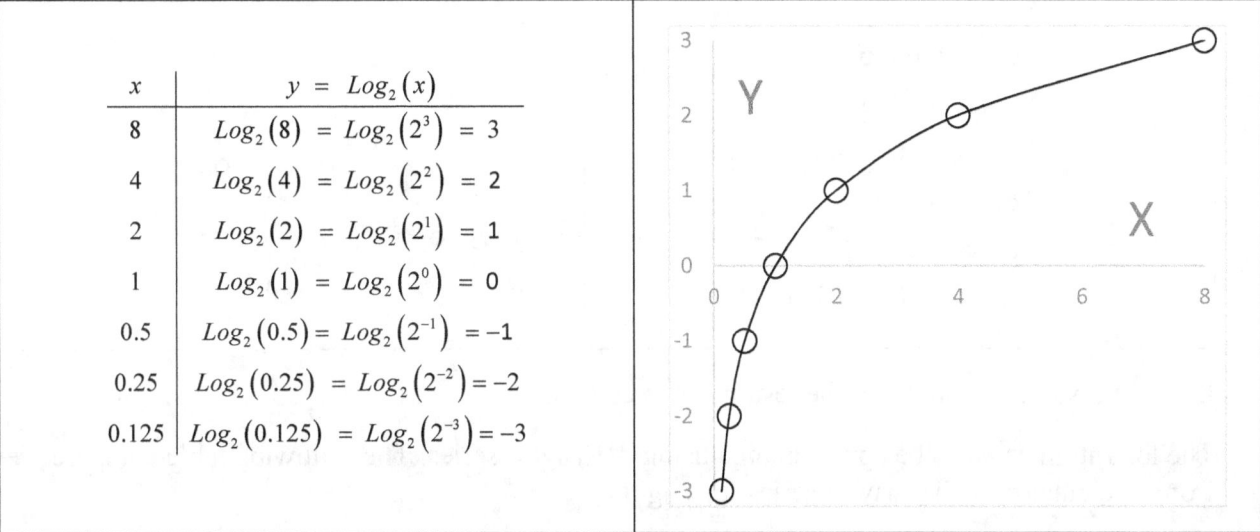

| $x$ | $y = Log_2(x)$ |
|---|---|
| 8 | $Log_2(8) = Log_2(2^3) = 3$ |
| 4 | $Log_2(4) = Log_2(2^2) = 2$ |
| 2 | $Log_2(2) = Log_2(2^1) = 1$ |
| 1 | $Log_2(1) = Log_2(2^0) = 0$ |
| 0.5 | $Log_2(0.5) = Log_2(2^{-1}) = -1$ |
| 0.25 | $Log_2(0.25) = Log_2(2^{-2}) = -2$ |
| 0.125 | $Log_2(0.125) = Log_2(2^{-3}) = -3$ |

Note: $Log_2(0)$ is undefined because $2^{-\infty}$ is never exactly zero.

How does a calculator do this for non-integers? The following is an introduction to this subject.

There is no specific formula for $2^x$. There are countless ways to do it. It is important to note that all of them are based on approximation. It is shown in calculus that the function $2^x$ can be approximated by the following Power series.

$$2^x = b_0 + b_1(x-a) + b_2(x-a)^2 + \bullet\bullet\bullet + b_n(x-a)^n$$

where a is an input constant and b is a computed constant. Theoretically, n approaches infinity. In practice however, n need only be large enough for the desired accuracy.

One way to compute the b's is called the Taylor series expansion. Let's compare the Taylor series with what a calculator would get. The following table shows the result when we use the first ten terms of the Taylor series for the function $2^x$.

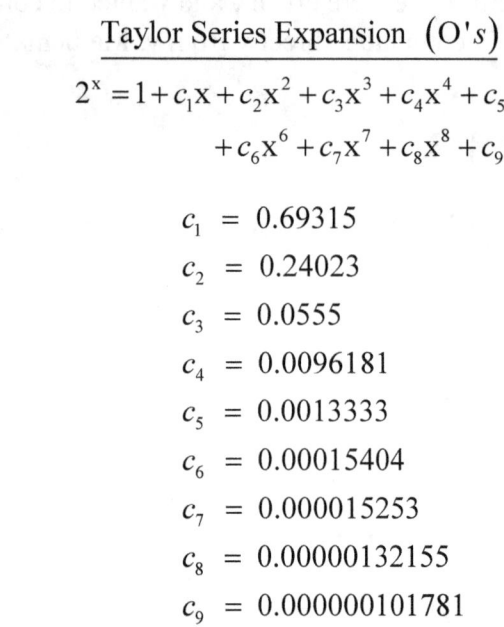

Taylor Series Expansion $(O's)$

$$2^x = 1 + c_1 x + c_2 x^2 + c_3 x^3 + c_4 x^4 + c_5 x^5$$
$$+ c_6 x^6 + c_7 x^7 + c_8 x^8 + c_9 x^9$$

$c_1 = 0.69315$

$c_2 = 0.24023$

$c_3 = 0.0555$

$c_4 = 0.0096181$

$c_5 = 0.0013333$

$c_6 = 0.00015404$

$c_7 = 0.000015253$

$c_8 = 0.00000132155$

$c_9 = 0.000000101781$

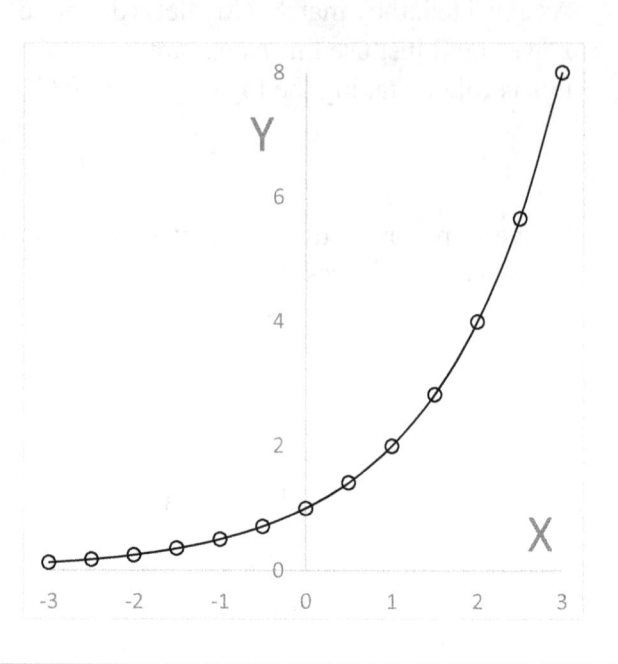

In the interval x = -3 to x = 3, the results are excellent.

The logarithm can also be approximated using the Taylor series. The following table compares results from a calculator, with a Taylor series for $Log_2(x)$.

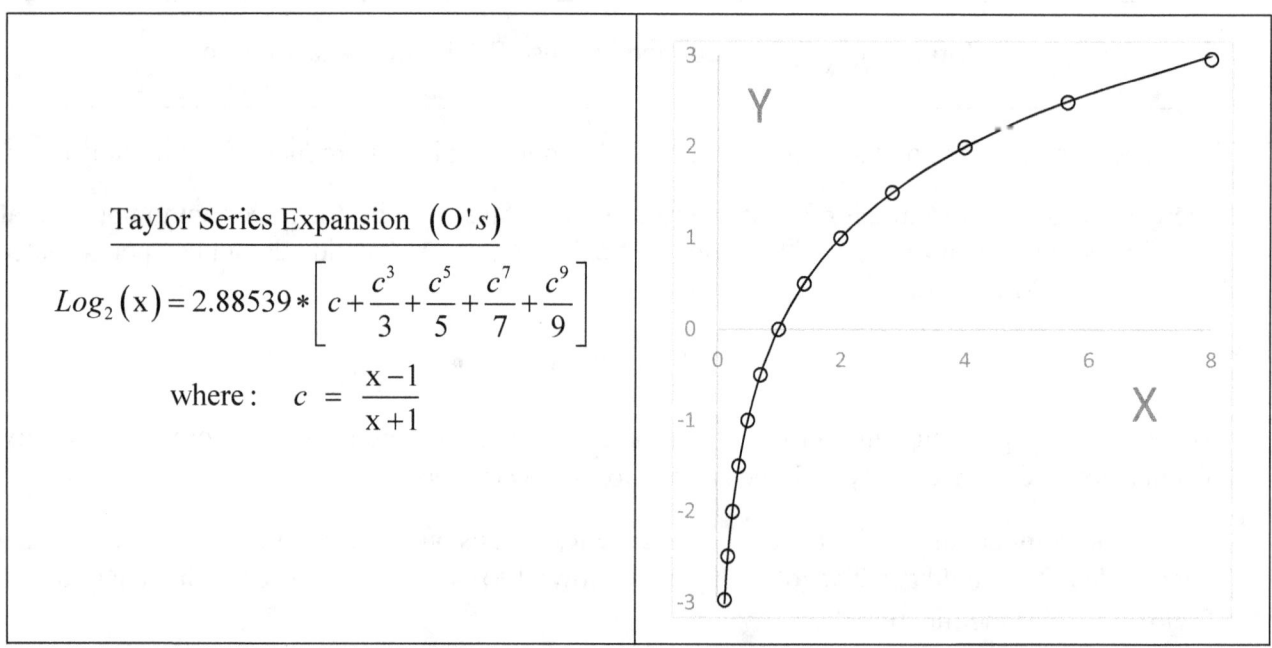

Taylor Series Expansion $(O's)$

$$Log_2(x) = 2.88539 * \left[ c + \frac{c^3}{3} + \frac{c^5}{5} + \frac{c^7}{7} + \frac{c^9}{9} \right]$$

$$\text{where:} \quad c = \frac{x-1}{x+1}$$

It is important to note that whatever method the calculator uses, it is also only an approximation.

## Base 10 Exponential and Logarithmic Functions

The equations for these are $y = 10^x$ and $y = \text{Log}_{10}(x)$. Obviously, this base is used because of our counting system. The following table shows a plot of these two functions.

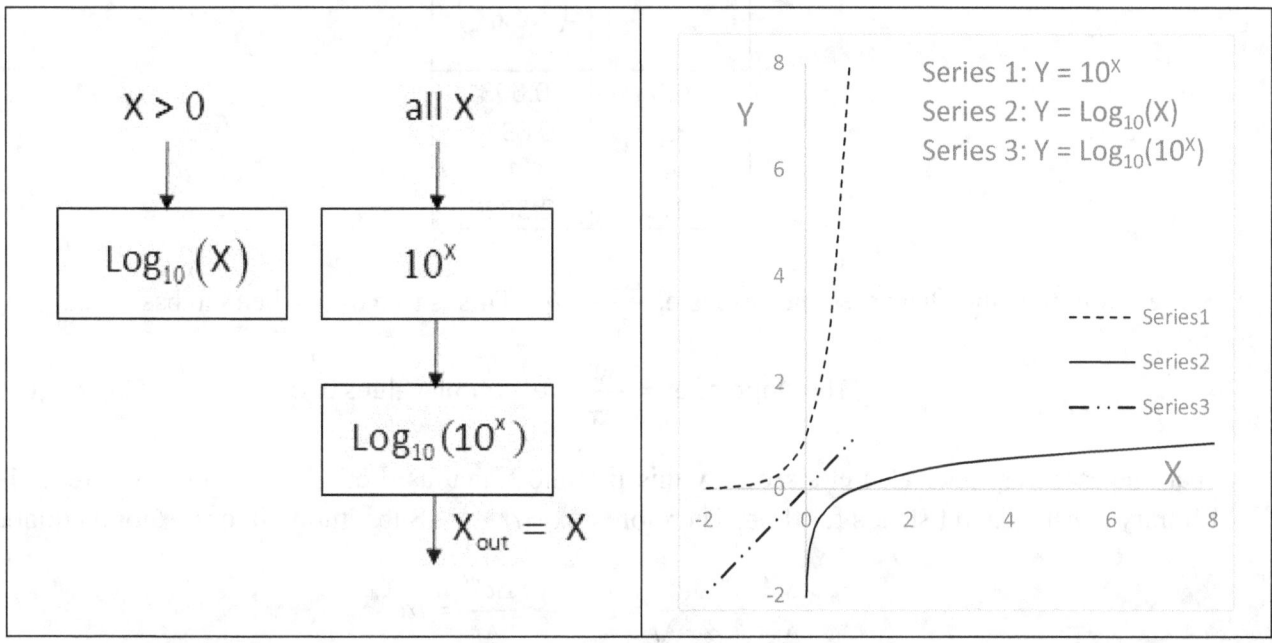

The block diagram emphasizes that $\text{Log}_{10} 10^x = x$. In fact, for any base, $\text{Log}_{base} base^x = x$. Also, for any base, $\text{Log}_{base}(0)$ is undefined. That's because $(base)^{-\infty}$ never gets to zero.

## Base e Exponential and Logarithmic Functions

Base e is so special that it has its own symbols.

$$y = e^x \quad \text{and} \quad y = \ln(x)$$

In these functions, e is an irrational number whose value is approximately 2.71828, and $\ln(x)$ stands for $\text{Log}_e(x)$.

Let's discuss why these are special functions. Calculus is a study of how things change. Change can be described by slope. Let's study the slope of the general exponential function $b^x$.

Define the change in x as $\Delta x$, and the change in $b^x$ as $\Delta b^x$. Then $\Delta b^x = b^{x+\Delta x} - b^x$. The slope of $b^x$ can be written

$$\frac{\Delta b^x}{\Delta x} = \frac{b^{x+\Delta x} - b^x}{\Delta x} = b^x \left( \frac{b^{\Delta x} - 1}{\Delta x} \right).$$

21

Let's evaluate the term $\left(\dfrac{b^{\Delta x}-1}{\Delta x}\right)$ for several values of b using $\Delta x = 0.0000001$.

| Base b | $\left(\dfrac{b^{\Delta x}-1}{\Delta x}\right)$ |
|:---:|:---:|
| 0.5 | -0.6931 |
| 2 | 0.6931 |
| e | 1 |
| 10 | 2.3026 |

We see from the table that when the base is e, $\dfrac{\Delta b^x}{\Delta x} = b^x$. This is true only when the base is e.

$$\text{The slope of } e^x = \frac{\Delta e^x}{\Delta x} = e^x \text{ for all values of } x \qquad \text{Slope Equation}$$

That makes e very special. Let's see how this fits into Calculus. Let's let $x = \alpha * t$, where $\alpha$ is an arbitrary constant and t stands for time. Therefore, $\Delta x = \alpha * \Delta t$. Substituting into the Slope Equation,

$$\frac{\Delta e^x}{\Delta x} = \frac{\Delta e^{\alpha t}}{\alpha * \Delta t} = e^{\alpha t} \;\; \rightarrow \;\; \frac{\Delta e^{\alpha t}}{\Delta t} = \alpha e^{\alpha t}$$

Suppose something moves with the formula $z = Ke^{\alpha t}$, where K is a constant. Then $\dfrac{\Delta\left(\dfrac{z}{K}\right)}{\Delta t} = \alpha\left(\dfrac{z}{K}\right)$. This

makes $\dfrac{\Delta z}{\Delta t} = \alpha z$. This calculus equation says that the rate of change of z is directly proportional to z itself. In school, this was the first differential equation I learned to solve. I learned that $z = Ke^{\alpha t}$ is the only possible solution, and it works only because of the slope equation.

Base e shows up everywhere in the study of motion. That's why ln is called the *natural* logarithm. The following table shows a plot of the base e exponential and logarithmic functions.

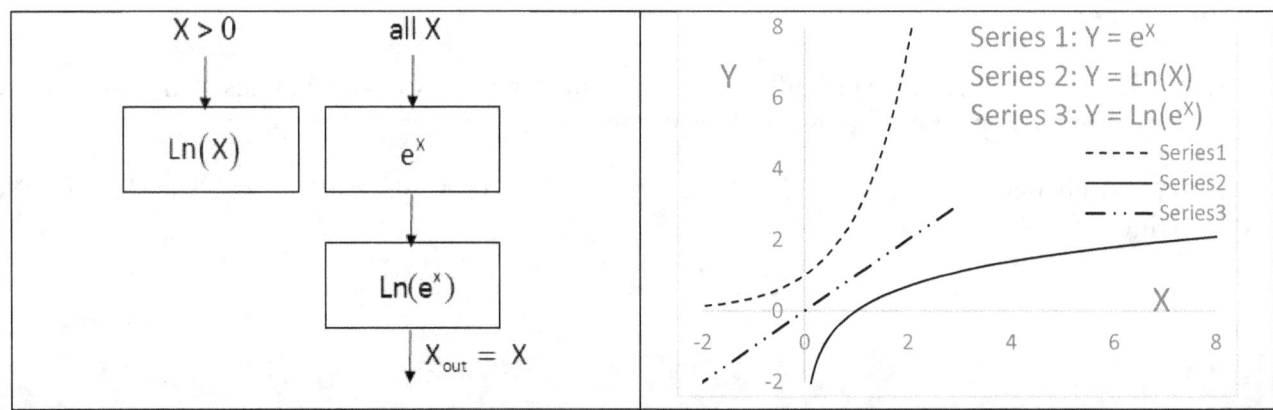

## Properties of the Exponential Function

The following table discusses important properties of the exponential function. These are valid for all positive bases and for all real exponents.

| The Property | Discussion |
|---|---|
| $b^x b^y = b^{x+y}$ | • This is easy to show using integers. $2^2 2^3 = (2*2)*(2*2*2) = 2^5 = 2^{2+3}$ <br><br> • The following is valid. $(1.9)^{2.4} * (1.9)^{1.6} = (1.9)^4$ |
| $\dfrac{b^x}{b^y} = b^x * b^{-y} = b^{x-y}$ | • If $x = y$, then $\dfrac{b^x}{b^x} = b^{x-x} = b^0 = 1$, which we see from previous plots. <br><br> • If $x = 0$, then $\dfrac{b^0}{b^y} = b^{0-y} = b^{-y} = \dfrac{1}{b^y}$. |
| $\left(b^x\right)^y = b^{x*y}$ | • Using integers, $\left(2^2\right)^3 = (2*2)*(2*2)*(2*2) = 2^6 = 2^{2*3}$. <br><br> • The following is valid. $\left((1.9)^{2.4}\right)^{\frac{5}{3}} = (1.9)^{2.4*\frac{5}{3}} = (1.9)^4$ |
| $(ab)^x = a^x b^x$ | • This is easy to show using integers. $(ab)^3 = (ab)*(ab)*(ab) = a^3 b^3$ <br><br> • The following is valid. $((1.6)(2.1))^{3.2} = (1.6)^{3.2}(2.1)^{3.2}$ |
| $\left(\dfrac{b}{a}\right)^{-x} = \left(\dfrac{a}{b}\right)^x$ | • $\left(\dfrac{b}{a}\right)^{-x} = \left(b*a^{-1}\right)^{-x} = b^{-x} a^x = \dfrac{a^x}{b^x} = \left(\dfrac{a}{b}\right)^x$. |
| $\left(\dfrac{a^x}{b^y}\right)^z = \dfrac{a^{xz}}{b^{yz}}$ | • This is a mix of several of the above properties. |

## Properties of the Logarithmic Function

From the basic definition, $\text{Log}_b b^x = x$. This yields $\text{Log}_b b = 1$, $\text{Log}_b(1) = 0$. Also, $\text{Log}_b(0)$ is undefined because $b^{-\infty}$ is never zero.

The following table shows important properties of the Log function.

| The Property | Discussion |
|---|---|
| (1) $b^{\log_b \alpha} = \alpha$ | If $\alpha = b^x$, then $\text{Log}_b(\alpha) = x$. Raising both sides to base b yields $$b^{\text{Log}_b(\alpha)} = b^x = \alpha.$$ |
| (2) $\text{Log}_b(\alpha\beta) = \text{Log}_b(\alpha) + \text{Log}_b(\beta)$ | Let $\text{Log}_b(\alpha) = u$. Raise both sides to base b. $b^{\text{Log}_b(\alpha)} = b^u$. From (1), $\alpha = b^u$. <br><br> Let $\text{Log}_b(\beta) = v$. Similarly, $\beta = b^v$. Therefore, $$\text{Log}_b(\alpha\beta) = \text{Log}_b(b^u b^v) = \text{Log}_b b^{u+v} = u + v = \text{Log}_b(\alpha) + \text{Log}_b(\beta)$$ Plug in numbers: $\text{Log}_2(8*4) = \text{Log}_2(8) + \text{Log}_2(4) = 3 + 2$ |
| (3) $\text{Log}_b\left(\dfrac{\alpha}{\beta}\right) = \text{Log}_b(\alpha) - \text{Log}_b(\beta)$ | Certainly $\text{Log}_b(\alpha) = \text{Log}_b\left(\dfrac{\alpha}{\beta}\beta\right)$ From (2), $$\text{Log}_b\left(\dfrac{\alpha}{\beta}\beta\right) = \text{Log}_b\left(\dfrac{\alpha}{\beta}\right) + \text{Log}_b(\beta) = \text{Log}_b(\alpha)$$ Rearranging, $$\text{Log}_b\left(\dfrac{\alpha}{\beta}\right) = \text{Log}_b(\alpha) - \text{Log}_b(\beta)$$ |
| (4) $\text{Log}_b(\alpha)^x = x * \text{Log}_b(\alpha)$ | Let $\text{Log}_b(\alpha) = u$. Raise both sides to base b. $b^{\text{Log}_b(\alpha)} = b^u$. From (1), $\alpha = b^u$. <br> Therefore, $$\text{Log}_b(\alpha)^x = \text{Log}_b(b^u)^x = \text{Log}_b b^{u*x} = u * x = x * \text{Log}_b(\alpha)$$ Plug in numbers: $\text{Log}_2 8^3 = 3 * \text{Log}_2 8$ |

## Logarithmic Change of Base

The following table shows how to compute $\text{Log}_6 15$.

| | |
|---|---|
| The equation to be solved is: | $x = \text{Log}_6 15$. |
| Raise both sides to base 6: | $6^x = 6^{\text{Log}_6 15}$. |
| Because $b^{\text{Log}_b \alpha} = \alpha$, | $6^x = 15$. |
| Take the natural log. | $\ln\left(6^x\right) = \ln(15)$ |
| Since $\text{Log}_b (\alpha)^x = x * \text{Log}_b (\alpha)$, | $x * \ln(6) = \ln(15)$. |
| Finally, | $x = \dfrac{\ln(15)}{\ln(6)}$. |

In general, $x = \text{Log}_{base} \alpha$ can be computed from $x = \dfrac{\ln(\alpha)}{\ln(base)}$ or $x = \dfrac{\text{Log}_{10}(\alpha)}{\text{Log}_{10}(base)}$ or other available Log functions.

The following table contains a few notes.

| |
|---|
| $\text{Log}_b (\alpha + \beta) \neq \text{Log}_b \alpha + \text{Log}_b \beta$ |
| $\text{Log}_b \dfrac{\alpha}{\beta} \neq \dfrac{\text{Log}_b \alpha}{\text{Log}_b \beta}$ |
| $\left(\text{Log}_b \alpha\right)^x \neq x * \text{Log}_b \alpha$ |

We will now discuss the VBA programming needed to do the math in this chapter.

# Exponential and Logarithmic Functions – Programming in Chapter 2

For the programs in this chapter, the following VBA syntax must be defined:

- The Array function;
- The statements Option Base 0 and Option Base 1.

The Array function is used to assign values to a variable. For example, alfa $=$ Array( -25.1, 50, 4).

• If Option Base 1 is the first card of the program, the index for the first value of alfa is 1. That means that alfa(1) = -25.1, alfa(2) = 50, and alfa(3) = 4.

• If Option Base 0 is the first card of the program, the index for the first value of alfa is 0. That means that alfa(0) = -25.1, alfa(1) = 50, and alfa(2) = 4. Note that Option Base 0 is the default.

The programs in the following table use the Option statement and produce the same output.

| | |
|---|---|
| Sub one()<br> alfa = Array(-25.1, 50, 4)<br> For j = 0 To 2<br>  Cells(1, j + 1) = alfa(j)<br> Next j<br>End Sub | • Option Base 0 is assumed. Therefore,<br>     alfa(0) = -25.1, alfa(1) = 50, alfa(2)=4<br><br>• When $j$ = 0, Cells(1, $j$ + 1) = Cells(1,1) |
| Option Base 1<br>Sub two()<br> alfa = Array(-25.1, 50, 4)<br> For j = 1 To 3<br>  Cells(1, J) = alfa(J)<br> Next j<br>End Sub | • Option Base 1 is declared. Therefore,<br>     alfa(1) = -25.1, alfa(2) = 50, alfa(3)=4 |
| This spreadsheet output is the same for both programs.<br><br>     -25.1     50     4 ||

We can now discuss the programs that were used in this chapter. The following table shows the block diagram of the first program. The data from the program is used to make the accompanying figure.

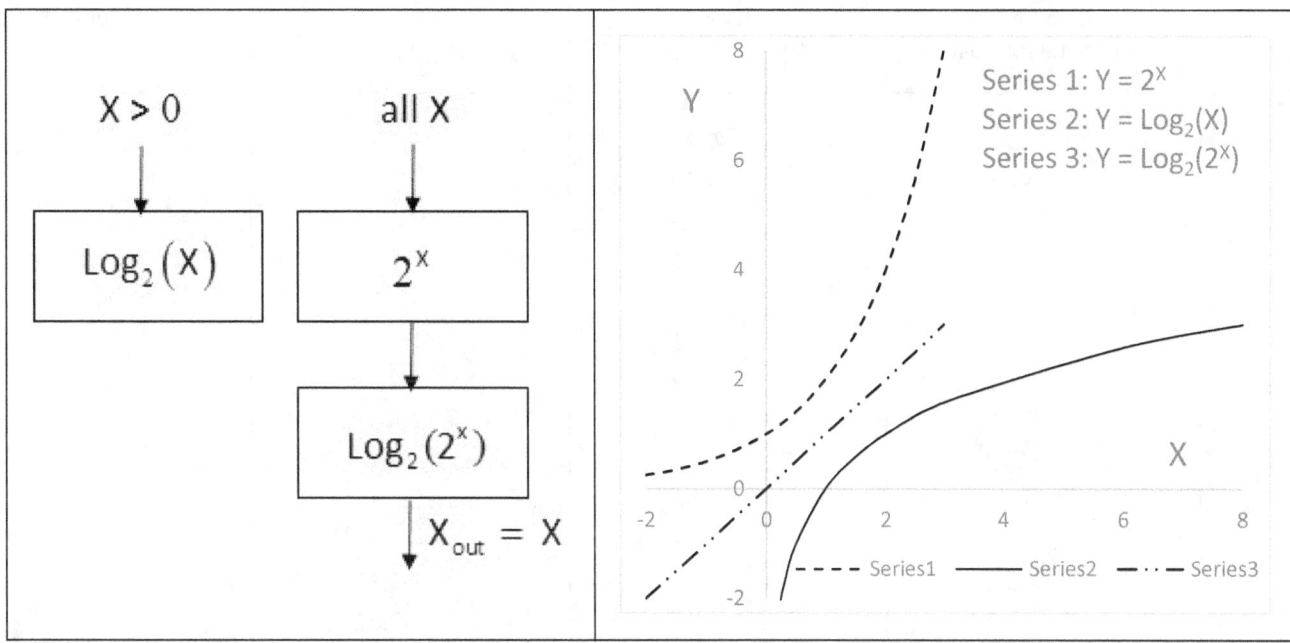

The following table is a listing of the program.

```
Option Base 1
Sub base_2()
  j = 1
For x = -2 To 3 Step 0.5
  Y = 2 ^ x
  Cells(j, 1) = x
  Cells(j, 2) = Application.Round(Y, 2)
  j = j + 1
Next x
Xa = Array(0.1, 0.2, 0.3, 0.5, 1, 1.5, 2, 3, 6, 8)
For j = 1 To 10
  Y = Application.Log(Xa(j), 2)
  Cells(j + 11, 1) = Xa(j)
  Cells(j + 11, 3) = Application.Round(Y, 2)
Next j
j = 1
For x = -2 To 3 Step 0.5
  Xout = Application.Log(2 ^ x, 2)
  Cells(j + 21, 1) = x
  Cells(j + 21, 4) = Application.Round(Xout, 2)
  j = j + 1
Next x
End Sub
```

• Syntax: $Log_2(x)$ is Application.Log(y, 2).
• Syntax for data roundoff is Application.Round(data, p).

• Each output series is put into its own column. This allows separate formatting when plotting.

• $Y = 2^X$ and is put in column 2 and is called Series 1.

• $Y = Log_2(x)$ and is put in column 3 and is called Series 2. This series gets the values of x from an Array function. In this way, x can be numbers with uneven intervals which is appropriate for the Log function.

• $Y = Log_2(2^X)$ and is put in column 4 and is called Series 3.

• Note that arithmetic can be done in the Cells statement.

The following table shows the spreadsheet from Program **Sub base_2.**

| Column 1<br>Horizontal Axis Data | Column 2<br>$Y = 2^X$ : Series 1 | Column 3<br>$Y = Log_2(X)$: Series 2 | Column 4<br>$Y = Log_2(2^X)$: Series 3 |
|:---:|:---:|:---:|:---:|
| -2 | 0.25 | | |
| -1.5 | 0.35 | | |
| -1 | 0.5 | | |
| -0.5 | 0.71 | | |
| 0 | 1 | | |
| 0.5 | 1.41 | | |
| 1 | 2 | | |
| 1.5 | 2.83 | | |
| 2 | 4 | | |
| 2.5 | 5.66 | | |
| 3 | 8 | | |
| 0.1 | | -3.32 | |
| 0.2 | | -2.32 | |
| 0.3 | | -1.74 | |
| 0.5 | | -1 | |
| 1 | | 0 | |
| 1.5 | | 0.58 | |
| 2 | | 1 | |
| 3 | | 1.58 | |
| 6 | | 2.58 | |
| 8 | | 3 | |
| -2 | | | -2 |
| -1.5 | | | -1.5 |
| -1 | | | -1 |
| -0.5 | | | -0.5 |
| 0 | | | 0 |
| 0.5 | | | 0.5 |
| 1 | | | 1 |
| 1.5 | | | 1.5 |
| 2 | | | 2 |
| 2.5 | | | 2.5 |
| 3 | | | 3 |

The rectangular array of spreadsheet cells that are selected for the Scatter chart: Cells(1, 1) to Cells(32,4).

The second program makes the following figure. It shows the comparison between the Taylor series and the VBA function for $2^x$.

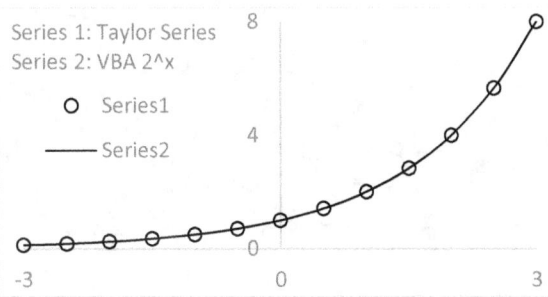

The following tables show the program and its output to the spreadsheet.

| | |
|---|---|
| ```Sub two_to_x()```<br>  ```c1 = 0.69315: c2 = 0.24023: c3 = 0.0555```<br>  ```c4 = 0.0096181: c5 = 0.0013333: c6 = 0.00015404```<br>  ```c7 = 0.000015253: c8 = 0.00000132155: c9 = 0.000000101781```<br>  ```i = 1```<br>  ```For x = -3 To 3 Step 0.5```<br>    ```two_x = 1 + c1 * x + c2 * x ^ 2 + c3 * x ^ 3 + c4 * x ^ 4 _```<br>    ```+ c5 * x ^ 5 + c6 * x ^ 6 + c7 * x ^ 7 + c8 * x ^ 8 + c9 * x ^ 9```<br>    ```Cells(i, 1) = x```<br>    ```Cells(i, 2) = Application.Round(two_x, 3)```<br>    ```Cells(i, 3) = Application.Round(2 ^ x, 3)```<br>    ```i = i + 1```<br>  ```Next x```<br>```End Sub``` | • A space followed by an underscore ( _ ) allows a line to continue on the next line.<br><br>• Scatter chart instructions in chapter 1 describe how to format a series using a line or markers. |

| Horizontal Axis Data | The Taylor Series for $2^x$ | The VBA Function for $2^x$ |
|---:|---:|---:|
| -3 | 0.125 | 0.125 |
| -2.5 | 0.177 | 0.177 |
| -2 | 0.25 | 0.25 |
| -1.5 | 0.354 | 0.354 |
| -1 | 0.5 | 0.5 |
| -0.5 | 0.707 | 0.707 |
| 0 | 1 | 1 |
| 0.5 | 1.414 | 1.414 |
| 1 | 2 | 2 |
| 1.5 | 2.828 | 2.828 |
| 2 | 4 | 4 |
| 2.5 | 5.657 | 5.657 |
| 3 | 7.999 | 8 |

The third program makes the following figure. It shows the comparison between the Taylor series and the VBA function for $Log_2(x)$.

The following tables show the program and its output to the spreadsheet.

| | |
|---|---|
| ```<br>Option Base 1<br>Sub Log2_x()<br>  ns = 13<br>  s = Array(0.125, 0.177, 0.25, 0.354, 0.5, 0.707, 1, _<br>          1.414, 2, 2.828, 4, 5.657, 8)<br>  For i = 1 To ns<br>    x = s(i)<br>    C = (x - 1) / (x + 1)<br>    TS_Log2 = 2.88539 * (C + C ^ 3 / 3 + C ^ 5 / 5 _<br>            + C ^ 7 / 7 + C ^ 9 / 9)<br>    XL_Log2 = Application.Log(x, 2)<br>    Cells(i, 1) = x<br>    Cells(i, 2) = Application.Round(TS_Log2, 3)<br>    Cells(i, 3) = Application.Round(XL_Log2, 3)<br>  Next i<br>End Sub<br>``` | • Option Base 1 makes the Array function begin with s(1).<br><br>• The Array function uses the uneven intervals that are required of a Log Function. In contrast, the For_Next loop uses even intervals. Step = 1 is the default. |

| Horizontal Axis Data | The Taylor Series for $Log_2(x)$ | The VBA Function for $Log_2(x)$ |
|---|---|---|
| 0.125 | -2.965 | -3 |
| 0.177 | -2.489 | -2.498 |
| 0.25 | -1.999 | -2 |
| 0.354 | -1.498 | -1.498 |
| 0.5 | -1 | -1 |
| 0.707 | -0.5 | -0.5 |
| 1 | 0 | 0 |
| 1.414 | 0.5 | 0.5 |
| 2 | 1 | 1 |
| 2.828 | 1.5 | 1.5 |
| 4 | 1.999 | 2 |
| 5.657 | 2.491 | 2.5 |
| 8 | 2.965 | 3 |

# Chapter 3: Equations and Matrices

The following is a typical algebra problem. We are given two equations:

$$v - 3w = 4$$
$$-2v + 4w = -6$$

We are asked to find the values for v and w that satisfy both equations together. One of the methods for doing this is the substitution method. Let's use this to solve our equations.

> Step 1: Solve the first equation for $v \rightarrow v = 4 + 3w$.
> Step 2: Substitute this into the second equation $\rightarrow -2*(4+3w)+4w = -6$.
> Step 3: Solve the last equation for $w \rightarrow w = -1$.
> Step 4: Substitute this into the Step 1 equation $\rightarrow v = 1$.
> Step 5: A check shows that these are correct.

In precalculus, we will change our equations into a matrix equation. This equation will have a standard form that fits all of the sets of linear equations.

## Creating a Matrix Equation

A matrix is a rectangular array of numbers. The following is a matrix that has two rows and one column.

$$\begin{bmatrix} a_{11} \\ a_{21} \end{bmatrix}$$

$a_{11}$ and $a_{21}$ represent numbers, and the indices give each a row-column location.

• Matrix equality: Two matrices are equal if their corresponding entries are equal. For example:

$$\text{If } \begin{bmatrix} a_{11} \\ a_{21} \end{bmatrix} = \begin{bmatrix} b_{11} \\ b_{21} \end{bmatrix}, \text{ then } a_{11} = b_{11} \text{ and } a_{21} = b_{21}.$$

Let's put our given system into matrix form. We will put the left-hand side into one matrix, and the right-hand side into another.

$$\begin{bmatrix} (v - 3w) \\ (-2v + 4w) \end{bmatrix} = \begin{bmatrix} 4 \\ -6 \end{bmatrix}$$

• Matrix multiplication: Two matrices can be multiplied only if the number of columns of the first matrix is equal to the number of rows of the second. For example:

If $C = \begin{bmatrix} a_{11} & a_{12} \end{bmatrix}$ and $D = \begin{bmatrix} b_{11} \\ b_{21} \end{bmatrix}$, then C*D exists. But $D*C$ does not. The following shows how C*D

is computed. $$C*D = \begin{bmatrix} a_{11} & a_{12} \end{bmatrix} * \begin{bmatrix} b_{11} \\ b_{21} \end{bmatrix} = \left[ \sum_{k=1}^{2} a_{1k} * b_{k1} \right] = \left[ (a_{11}*b_{11} + a_{12}*b_{21}) \right]$$

The matrix $C*D$ has one row and one column. It has the same number of rows as C, and the same number of columns as D. The matrix $C*D$ is a *sum of products*. Each product has this form:

$$a_{1k}*b_{k1}$$

Important Note: k is the index for the columns of the first matrix. It is the same index that is used for the rows of the second. For $C*D$, k runs from 1 to 2.

Let's add a row to matrix C.

$$C = \begin{bmatrix} a_{11} & a_{12} \\ a_{21} & a_{22} \end{bmatrix}$$

If we multiply this by the same D matrix, the second row is dealt with just like the first.

$$C*D = \begin{bmatrix} a_{11} & a_{12} \\ a_{21} & a_{22} \end{bmatrix} * \begin{bmatrix} b_{11} \\ b_{21} \end{bmatrix} = \begin{bmatrix} \sum_{k=1}^{2} a_{1k}*b_{k1} \\ \sum_{k=1}^{2} a_{2k}*b_{k1} \end{bmatrix} = \begin{bmatrix} \left(a_{11}*b_{11} + a_{12}*b_{21}\right) \\ \left(a_{21}*b_{11} + a_{22}*b_{21}\right) \end{bmatrix}$$

For this case, $C*D$ has two rows and one column. Again, it has the same number of rows as C and the same number of columns as D.

Let's repeat our original system.

$$\begin{bmatrix} (v-3w) \\ (-2v+4w) \end{bmatrix} = \begin{bmatrix} 4 \\ -6 \end{bmatrix}$$

On the left-hand side, let's put the coefficients into a separate matrix: $A = \begin{bmatrix} 1 & -3 \\ -2 & 4 \end{bmatrix}$

Also, let's put the variables into a separate matrix: $x = \begin{bmatrix} v \\ w \end{bmatrix}$

The left-hand side can now be written as the product of two matrices: $A*x$

Finally, let's name the matrix of constants on the right-hand side: $b = \begin{bmatrix} 4 \\ -6 \end{bmatrix}$

Our system can now be written as $A*x=b$. There are two cases:

- Given A and b, find x.
- Given A and x, find b.

## Given A and b, Find x

We repeat our system in matrix form:

$$\begin{bmatrix} 1 & -3 \\ -2 & 4 \end{bmatrix} \begin{bmatrix} x_1 \\ x_2 \end{bmatrix} = \begin{bmatrix} 4 \\ -6 \end{bmatrix}$$

We begin by discussing three things we can do to the equations without affecting the solution. These things are called elementary row operations.

- Row operation #1: Interchanging equations (rows) doesn't affect the answer.
- Row operation #2: Multiplying one of the rows by a non-zero constant doesn't affect the answer.
- Row operation #3: Adding or subtracting one row with another doesn't affect the answer.

We will use these operations to change our system into the following form:

$$\begin{bmatrix} 1 & 0 \\ 0 & 1 \end{bmatrix} \begin{bmatrix} x_1 \\ x_2 \end{bmatrix} = \begin{bmatrix} 1 \\ -1 \end{bmatrix}$$

The coefficient matrix has been diagonalized, and the solution is exposed. Let's demonstrate this in an example. For shorthand, we will only show the constants.

| Example lineq 1: Solve these equations. | $\begin{bmatrix} 1 & -3 \\ -2 & 4 \end{bmatrix} \begin{bmatrix} x_1 \\ x_2 \end{bmatrix} = \begin{bmatrix} 4 \\ -6 \end{bmatrix}$ | | |
|---|---|---|---|
| The initial matrices | 1<br>-2 | -3<br>4 | 4<br>-6 |
| If we multiply row 1 by 2, and add the result to row 2, we get a new row 2. | 1<br>0 | -3<br>-2 | 4<br>2 |
| Multiply row 2 by 3/2 and subtract result from row 1. We get a new row 1. | 1<br>0 | 0<br>-2 | 1<br>2 |
| Now divide row 2 by (−2) → | 1<br>0 | 0<br>1 | 1<br>-1 |

This agrees with the substitution method.

Example lineq 2: Solve these equations: $\begin{bmatrix} 1 & 0 & 1 \\ 2 & 3 & 1 \\ 3 & 5 & 2 \end{bmatrix} \begin{bmatrix} x_1 \\ x_2 \\ x_3 \end{bmatrix} = \begin{bmatrix} 4 \\ 8 \\ 6 \end{bmatrix}$.

Let's use the following notation. When zeroing elements below the diagonal, the steps will be labeled 1.1, 1.2, and so forth. For elements above the diagonal, the steps will be 2.1, 2.2, and so forth.

| Example lineq 2 | | | | |
|---|---|---|---|---|
| The initial matrices | 1<br>2<br>3 | 0<br>3<br>5 | 1<br>1<br>2 | 4<br>8<br>6 |
| Step 1.1: Referring to the initial matrices, the goal is to force $a_{21} = 0$. The multiplier is $a_{21}/a_{11}$. Multiply row 1 by 2, and subtract it from row 2. | 1<br>0<br>3 | 0<br>3<br>5 | 1<br>-1<br>2 | 4<br>0<br>6 |
| Step 1.2: Referring to the matrix of Step 1.1, the goal is to force $a_{31} = 0$. The multiplier is $a_{31}/a_{11}$. Multiply row 1 by 3, and subtract it from row 3. | 1<br>0<br>0 | 0<br>3<br>5 | 1<br>-1<br>-1 | 4<br>0<br>-6 |
| Step 1.3: Referring to the matrix of Step 1.2, the goal is to force $a_{32} = 0$. The multiplier is $a_{32}/a_{22}$. Multiply row 2 by 5/3, and subtract it from row 3. | 1<br>0<br>0 | 0<br>3<br>0 | 1<br>-1<br>2/3 | 4<br>0<br>-6 |
| Step 2.1: Referring to the last matrix, the goal is to force $a_{13} = 0$. The multiplier is $a_{13}/a_{33}$. Multiply row 3 by 3/2, and subtract it from row 1. | 1<br>0<br>0 | 0<br>3<br>0 | 0<br>-1<br>2/3 | 13<br>0<br>-6 |
| Step 2.2: Referring to the last matrix, the goal is to force $a_{23} = 0$. The multiplier is $a_{23}/a_{33}$. Multiply row 3 by (−3/2), and subtract it from row 2. | 1<br>0<br>0 | 0<br>3<br>0 | 0<br>0<br>2/3 | 13<br>-9<br>-6 |
| Step 2.3 Referring to the last matrix, $a_{12} = 0$. Therefore, the process is complete. To reveal the solution, unitize the diagonal of the first three columns. | 1<br>0<br>0 | 0<br>1<br>0 | 0<br>0<br>1 | 13<br>-3<br>-9 |

There is a pattern in these steps. When the element in any column p is being zeroed, each element in that row is modified by the following equation:

$$A_{i,j} = A_{i,j} - A_{p,j}\frac{A_{i,p}}{A_{p,p}}$$  Equation Step

Let's study **Equation Step**. At row $i = p+2$ and column $j = p$:

$$A_{p+2,\,p} = A_{p+2,\,p} - A_{p,\,p}\frac{A_{p+2,\,p}}{A_{p,\,p}} = 0.$$

An element below the diagonal element has been zeroed. Because all of the elements in row $i = p+2$ are being modified by **equation Step**, the solution has not been changed. Equation Step also applies to the rows above the diagonal.

To avoid $A_{p,p} = 0$, rows must be interchanged. This is shown in the following example.

Example lineq 3: Solve these equations: $\begin{bmatrix} 0 & 0 & 1 \\ 2 & -3 & 4 \\ 3 & 5 & -1 \end{bmatrix} \begin{bmatrix} x_1 \\ x_2 \\ x_3 \end{bmatrix} = \begin{bmatrix} 5 \\ 4 \\ -10 \end{bmatrix}$.

| Example lineq 3 | | | | |
|---|---|---|---|---|
| The initial matrices | 0 | 0 | 1 | 5 |
| | 2 | -3 | 4 | 4 |
| | 3 | 5 | -1 | -10 |
| Step 1.1: Referring to the initial matrices, the goal is to force $a_{21} = 0$. The multiplier is $a_{21}/a_{11}$. Because $a_{11} = 0$, interchange rows 1 and 2. | 2 | -3 | 4 | 4 |
| | 0 | 0 | 1 | 5 |
| | 3 | 5 | -1 | -10 |
| Step 1.2: Referring to the matrix of Step 1.1, the goal is to force $a_{31} = 0$. The multiplier is $a_{31}/a_{11}$. Multiply row 1 by 3/2, and subtract it from row 3. | 2 | -3 | 4 | 4 |
| | 0 | 0 | 1 | 5 |
| | 0 | 9.5 | -7 | -16 |
| Step 1.3: Referring to the matrix of Step 1.2, the goal is to force $a_{32} = 0$. The multiplier is $a_{32}/a_{22}$. Because $a_{22} = 0$, interchange rows 2 and 3. | 2 | -3 | 4 | 4 |
| | 0 | 9.5 | -7 | -16 |
| | 0 | 0 | 1 | 5 |
| Step 2.1: Referring to the matrix of Step 1.3, the goal is to force $a_{13} = 0$. The multiplier is $a_{13}/a_{33}$. Multiply row 3 by 4, and subtract it from row 1. | 2 | -3 | 0 | -16 |
| | 0 | 9.5 | -7 | -16 |
| | 0 | 0 | 1 | 5 |
| Step 2.2: Referring to the matrix of Step 2.1, the goal is to force $a_{23} = 0$. The multiplier is $a_{23}/a_{33}$. Multiply row 3 by (-7), and subtract it from row 2. | 2 | -3 | 0 | -16 |
| | 0 | 9.5 | 0 | 19 |
| | 0 | 0 | 1 | 5 |
| Step 2.3: Referring to the matrix of Step 2.2, the goal is to force $a_{12} = 0$. The multiplier is $a_{12}/a_{22}$. Multiply row 2 by (-3/9.5) and subtract it from row 1. | 2 | 0 | 0 | -10 |
| | 0 | 9.5 | 0 | 19 |
| | 0 | 0 | 1 | 5 |
| Referring to the matrix of Step 2.3, unitize the diagonal of the first three columns. | 1 | 0 | 0 | -5 |
| | 0 | 1 | 0 | 2 |
| | 0 | 0 | 1 | 5 |

Example lineq 4: Solve these equations: $\begin{bmatrix} 4 & 2 & -4 \\ 2 & 1 & -1 \\ 6 & 3 & -5 \end{bmatrix} \begin{bmatrix} x_1 \\ x_2 \\ x_3 \end{bmatrix} = \begin{bmatrix} 6 \\ 4 \\ 6 \end{bmatrix}$.

| Example lineq 4 | | | | |
|---|---|---|---|---|
| The initial matrices. | 4 | 2 | -4 | 6 |
| | 2 | 1 | -1 | 4 |
| | 6 | 3 | -5 | 6 |
| Step 1.1: Referring to the initial matrices, the goal is to force $a_{21}=0$. The multiplier is $a_{21}/a_{11}$. Multiply row 1 by 1/2, and subtract it from row 2. | 4 | 2 | -4 | 6 |
| | 0 | 0 | 1 | 1 |
| | 6 | 3 | -5 | 6 |
| Step 1.2: Referring to the matrix in Step 1.1, the goal is to force $a_{31} = 0$. The multiplier is $a_{31}/a_{11}$. Multiply row 1 by 3/2 and subtract it from row 3. | 4 | 2 | -4 | 6 |
| | 0 | 0 | 1 | 1 |
| | 0 | 0 | 1 | -3 |
| Step 1.3: Referring to the matrix in Step 1.2, the goal is to force $a_{32} = 0$. The multiplier is $a_{32}/a_{22}$. Since it is not possible to interchange rows, there is no solution. | There is no solution. | | | |
| The original equations 1 and 3 require that $4x_1 + 2x_2 - 4x_3 = 6x_1 + 3x_2 - 5x_3$, or $2x_1 + x_2 - x_3 = 0$. This contradicts Equation 2. The original equations are inconsistent and they have no solution. | | | | |

## Matrices

A matrix is a rectangular array of numbers, arranged in M rows and N columns. It is customary to contain the array in brackets.

The following is a constant matrix.

$$A = \begin{bmatrix} 1 & 0 & 1 \\ 2 & 3 & 1 \end{bmatrix}$$

Having two rows and three columns, its size is identified as (2 x 3). Let's give each element a name with subscripts that identify its row-column location.

$$A = \begin{bmatrix} a_{11} & a_{12} & a_{13} \\ a_{21} & a_{22} & a_{23} \end{bmatrix}$$
$$(2 \times 3) = (M \times N)$$

1. Matrices are equal if their elements are equal.

We are given: $A = \underbrace{\begin{bmatrix} a_{11} & a_{12} & a_{13} \end{bmatrix}}_{(1 \times 3)}$ and $B = \underbrace{\begin{bmatrix} b_{11} & b_{12} & b_{13} \end{bmatrix}}_{(1 \times 3)}$.

Then $A = B$ if $a_{11} = b_{11}$, $a_{12} = b_{12}$, and $a_{13} = b_{13}$.

2. Matrices can be added or subtracted. For the A and B above,

$$A + B = \underbrace{\begin{bmatrix} (a_{11} + b_{11}) & (a_{12} + b_{12}) & (a_{13} + b_{13}) \end{bmatrix}}_{(1 \times 3)}$$

3. When a matrix is multiplied by a constant, each element is multiplied.

If $A = \underbrace{\begin{bmatrix} a_{11} & a_{12} \end{bmatrix}}_{(1 \times 2)}$, then $(-2) * A = \underbrace{\begin{bmatrix} -2a_{11} & -2a_{12} \end{bmatrix}}_{(1 \times 2)}$

4. A matrix can be multiplied by another matrix. The result is a matrix. The following shows the row-column relationship.

$$\underset{M \times L}{A} * \underset{L \times N}{B} = \underset{M \times N}{C}$$

- The number of columns of A must equal the number of rows of B.
- The number of rows of the product equals the number of rows of the first matrix.
- The number of columns of the product equals the number of columns of the second.

Each of the M*N elements of C is formed by the equation $C_{ij} = \sum\limits_{k=1}^{L} a_{ik} * b_{kj}$.

If $A = \begin{bmatrix} a_{11} & a_{12} & a_{13} \end{bmatrix}$ and $B = \begin{bmatrix} b_{11} \\ b_{21} \\ b_{31} \end{bmatrix}$, then A and B can be multiplied, and their product will have one

$\underbrace{\phantom{aaaaa}}_{(1 \times 3)}$ $\underbrace{\phantom{a}}_{(3 \times 1)}$

row and one column. Computing this product is rather straightforward.

$$C = A*B = \underbrace{\begin{bmatrix} a_{11} & a_{12} & a_{13} \end{bmatrix}}_{1 \times 3} * \underbrace{\begin{bmatrix} b_{11} \\ b_{21} \\ b_{31} \end{bmatrix}}_{3 \times 1} = \begin{bmatrix} \sum_{k=1}^{3} a_{1k}*b_{k1} \end{bmatrix} = \underbrace{\begin{bmatrix} a_{11}*b_{11} + a_{12}*b_{21} + a_{13}*b_{31} \end{bmatrix}}_{1 \times 1} = \underbrace{\begin{bmatrix} c_{11} \end{bmatrix}}_{1 \times 1}$$

Plugging in numbers, if $A = \underbrace{\begin{bmatrix} 1 & 2 & 3 \end{bmatrix}}_{1 \times 3}$ and $B = \underbrace{\begin{bmatrix} 5 \\ 6 \\ -7 \end{bmatrix}}_{3 \times 1}$, then $C = \underbrace{\begin{bmatrix} 1*5 + 2*6 - 3*7 \end{bmatrix}}_{1 \times 1} = \underbrace{\begin{bmatrix} -4 \end{bmatrix}}_{1 \times 1}$.

**Example mult.1:** If another row is added to A, the second row is treated just like the first.

If $A = \underbrace{\begin{bmatrix} a_{11} & a_{12} & a_{13} \\ a_{21} & a_{22} & a_{23} \end{bmatrix}}_{(2 \times 3)}$ and $B = \underbrace{\begin{bmatrix} b_{11} \\ b_{21} \\ b_{31} \end{bmatrix}}_{(3 \times 1)}$, then

$$C = A*B = \underbrace{\begin{bmatrix} a_{11} & a_{12} & a_{13} \\ a_{21} & a_{22} & a_{23} \end{bmatrix}}_{(2 \times 3)} * \underbrace{\begin{bmatrix} b_{11} \\ b_{21} \\ b_{31} \end{bmatrix}}_{(3 \times 1)} = \begin{bmatrix} \sum_{k=1}^{3} a_{1k}*b_{k1} \\ \sum_{k=1}^{3} a_{2k}*b_{k1} \end{bmatrix} = \underbrace{\begin{bmatrix} a_{11}b_{11} + a_{12}b_{21} + a_{13}b_{31} \\ a_{21}b_{11} + a_{22}b_{21} + a_{23}b_{31} \end{bmatrix}}_{(2 \times 1)} = \underbrace{\begin{bmatrix} c_{11} \\ c_{21} \end{bmatrix}}_{(2 \times 1)}.$$

Plugging in numbers, if $A = \underbrace{\begin{bmatrix} 1 & 2 & 3 \\ -1 & -2 & -3 \end{bmatrix}}_{(2 \times 3)}$ and $B = \underbrace{\begin{bmatrix} 5 \\ 6 \\ -7 \end{bmatrix}}_{(3 \times 1)}$, then

$$C = A*B = \underbrace{\begin{bmatrix} 1 & 2 & 3 \\ -1 & -2 & -3 \end{bmatrix}}_{(2 \times 3)} * \underbrace{\begin{bmatrix} 5 \\ 6 \\ -7 \end{bmatrix}}_{(3 \times 1)} = \underbrace{\begin{bmatrix} 5+12-21 \\ -5-12+21 \end{bmatrix}}_{(2 \times 1)} = \underbrace{\begin{bmatrix} -4 \\ 4 \end{bmatrix}}_{(2 \times 1)}.$$

**Example mult.2:**

If $A=\begin{bmatrix} a_{11} & a_{12} & a_{13} \end{bmatrix}$ (1x3) and $B=\begin{bmatrix} b_{11} & b_{12} \\ b_{21} & b_{22} \\ b_{31} & b_{32} \end{bmatrix}$ (3x2), then $A*B = C$, where

$$C = \begin{bmatrix} \sum_{k=1}^{3} a_{1k}*b_{k1} & \sum_{k=1}^{3} a_{1k}*b_{k2} \end{bmatrix} = \begin{bmatrix} (a_{11}b_{11}+a_{12}b_{21}+a_{13}b_{31}) & (a_{11}b_{12}+a_{12}b_{22}+a_{13}b_{32}) \end{bmatrix}_{(1x2)} = \begin{bmatrix} c_{11} & c_{12} \end{bmatrix}_{(1x2)}.$$

Plugging in numbers, if $A=\begin{bmatrix} 1 & 2 & 3 \end{bmatrix}$ (1x3) and $B=\begin{bmatrix} 5 & 5 \\ 6 & 6 \\ -7 & -7 \end{bmatrix}$ 3x2, then

$$C = A*B = \begin{bmatrix} 1 & 2 & 3 \end{bmatrix}_{(1x3)} * \begin{bmatrix} 5 & 5 \\ 6 & 6 \\ -7 & -7 \end{bmatrix}_{(3x2)} = \begin{bmatrix} (5+12-21) & (5+12-21) \end{bmatrix}_{(1x2)} = \begin{bmatrix} -4 & -4 \end{bmatrix}_{(1x2)}.$$

**Example mult.3:**

If $A=\begin{bmatrix} a_{11} & a_{12} & a_{13} \\ a_{21} & a_{22} & a_{23} \end{bmatrix}$ (2x3) and $B=\begin{bmatrix} b_{11} & b_{12} \\ b_{21} & b_{22} \\ b_{31} & b_{32} \end{bmatrix}$ (3x2), then $A*B = C$, where

$$C = \begin{bmatrix} \sum_{k=1}^{3} a_{1k}*b_{k1} & \sum_{k=1}^{3} a_{1k}*b_{k2} \\ \sum_{k=1}^{3} a_{2k}*b_{k1} & \sum_{k=1}^{3} a_{2k}*b_{k2} \end{bmatrix} = \begin{bmatrix} (a_{11}b_{11}+a_{12}b_{21}+a_{13}b_{31}) & (a_{11}b_{12}+a_{12}b_{22}+a_{13}b_{32}) \\ (a_{21}b_{11}+a_{22}b_{21}+a_{23}b_{31}) & (a_{21}b_{12}+a_{22}b_{22}+a_{23}b_{32}) \end{bmatrix}_{(2x2)} = \begin{bmatrix} c_{11} & c_{12} \\ c_{21} & c_{22} \end{bmatrix}_{(2x2)}.$$

Plugging in numbers, if $A=\begin{bmatrix} 1 & 2 & 3 \\ -1 & -2 & -3 \end{bmatrix}$ (2x3) and $B=\begin{bmatrix} 5 & 5 \\ 6 & 6 \\ -7 & -7 \end{bmatrix}$ 3x2, then $A*B = C$, where

$$C = \begin{bmatrix} 1 & 2 & 3 \\ -1 & -2 & -3 \end{bmatrix}_{(2x3)} * \begin{bmatrix} 5 & 5 \\ 6 & 6 \\ -7 & -7 \end{bmatrix}_{(3x2)} = \begin{bmatrix} (5+12-21) & (5+12-21) \\ (-5-12+21) & (-5-12+21) \end{bmatrix}_{(2x2)} = \begin{bmatrix} -4 & -4 \\ 4 & 4 \end{bmatrix}_{(2x2)} = \begin{bmatrix} c_{11} & c_{12} \\ c_{21} & c_{22} \end{bmatrix}_{(2x2)}.$$

Because of the dimensions of A and B, they can also be multiplied in the reverse order.

$$C = \begin{bmatrix} 5 & 5 \\ 6 & 6 \\ -7 & -7 \end{bmatrix} * \begin{bmatrix} 1 & 2 & 3 \\ -1 & -2 & -3 \end{bmatrix} = \begin{bmatrix} (5-5) & (10-10) & (15-15) \\ (6-6) & (12-12) & (18-18) \\ (-7+7) & (-14+14) & (-21+21) \end{bmatrix} = \begin{bmatrix} 0 & 0 & 0 \\ 0 & 0 & 0 \\ 0 & 0 & 0 \end{bmatrix}$$

$\underbrace{\qquad}_{(3 \times 2)}$ $\underbrace{\qquad}_{(2 \times 3)}$ $\underbrace{\qquad}_{(3 \times 3)}$ $\underbrace{\qquad}_{(3 \times 3)}$

This clearly shows that A\*B is not necessarily equal to B\*A.

Example mult.4: If Ax = b, then if we are given A and x, we can find b.

$$\text{If} \quad A = \begin{bmatrix} 1 & 0 & 1 \\ 2 & 3 & 1 \\ 3 & 5 & 2 \end{bmatrix} \quad \text{and} \quad x = \begin{bmatrix} 13 \\ -3 \\ -9 \end{bmatrix}, \text{ then}$$

$\underbrace{\qquad}_{(3 \times 3)}$ $\underbrace{\qquad}_{3 \times 1}$

$$b = A * x = \begin{bmatrix} 1 & 0 & 1 \\ 2 & 3 & 1 \\ 3 & 5 & 2 \end{bmatrix} \begin{bmatrix} 13 \\ -3 \\ -9 \end{bmatrix} = \begin{bmatrix} (13-9) \\ (26-9-9) \\ (39-15-18) \end{bmatrix} = \begin{bmatrix} 4 \\ 8 \\ 6 \end{bmatrix}.$$

$\underbrace{\qquad}_{(3 \times 3)}$ $\underbrace{\qquad}_{3 \times 1}$ $\underbrace{\qquad}_{3 \times 1}$ $\underbrace{\qquad}_{3 \times 1}$

This verifies Example lineq 2.

Example mult.5:

$$\text{If} \quad A = \begin{bmatrix} 1 & 1/2 & 1/3 \\ 1/2 & 1/3 & 1/4 \\ 1/3 & 1/4 & 1/5 \end{bmatrix} \quad \text{and} \quad S = \begin{bmatrix} 9 & 36 & 30 \\ -36 & 192 & -180 \\ 30 & -180 & 180 \end{bmatrix}, \text{ then } A*S = C, \text{ where}$$

$\underbrace{\qquad}_{(3 \times 3)}$ $\underbrace{\qquad}_{(3 \times 3)}$

$$C = \begin{bmatrix} (9-18+10) & (-36+96-60) & (30-90+60) \\ (9/2-12+30/4) & (-18+64-45) & (15-60+45) \\ (3-9+6) & (-12+48-36) & (10-45+36) \end{bmatrix} = \begin{bmatrix} 1 & 0 & 0 \\ 0 & 1 & 0 \\ 0 & 0 & 1 \end{bmatrix}.$$

$\underbrace{\qquad}_{(3 \times 3)}$ $\underbrace{\qquad}_{(3 \times 3)}$

The elements of C are all zero except for the ones on its diagonal. Such a matrix is very important.

It is named the *Identity* matrix, and in general, is written as:

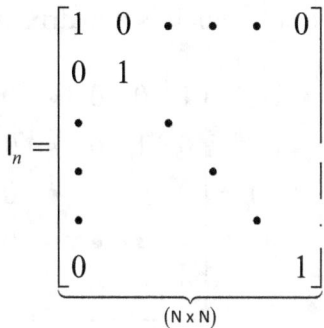

$$I_n = \begin{bmatrix} 1 & 0 & \cdot & \cdot & \cdot & 0 \\ 0 & 1 & & & & \\ \cdot & & \cdot & & & \\ \cdot & & & \cdot & & \\ \cdot & & & & \cdot & \\ 0 & & & & & 1 \end{bmatrix}$$
$$(N \times N)$$

Note: If G is an N x M matrix, then $\underset{N \times N}{I_n} * \underset{N \times M}{G} = \underset{N \times M}{G}$. Note that $\underset{N \times M}{G} * \underset{N \times N}{I_n}$ does not exist.

For this example, $A*S = I_n$. For this to be true, S must be the inverse of A. Indeed it is, and that's the definition of an inverse of a matrix. Matrix $A^{-1}$ is the inverse of matrix A if

$$A * A^{-1} = A^{-1} * A = I_n$$

5. For a square matrix, division is achieved by multiplication by its inverse. For example, suppose we want x when we are given

$$\underset{N \times N}{A} * \underset{N \times 1}{x} = \underset{N \times 1}{b}$$

If we multiply this equation by $A^{-1}$, we get $A^{-1}*A*x = A^{-1}*b$, or $x = A^{-1}*b$. Note that this is another way to solve for x.

In the following section, we will learn how to compute $A^{-1}$.

## Computing the Inverse of a Matrix

We begin with the identity matrix ( $I_n$ ), which is a matrix containing all zeros except for the ones on its diagonal.

$$I_n = \begin{bmatrix} 1 & 0 & 0 & \bullet & 0 \\ 0 & 1 & 0 & \bullet & 0 \\ 0 & 0 & 1 & \bullet & 0 \\ \bullet & \bullet & \bullet & \bullet & 0 \\ 0 & 0 & 0 & \bullet & 1 \end{bmatrix}.$$

$$\underbrace{\phantom{xxxxx}}_{(n \times n)}$$

The matrix $A^{-1}$ is the inverse of A, if $A^{-1} * A = A * A^{-1} = I_n$.

We started the chapter with the equation $Ax = b$. We formed the matrix $[\; A \; \vdots \; b \;]$. We then used *row operations* to change it to $[\; I_n \; \vdots \; x \;]$. We can do the same thing to compute $A^{-1}$. Form the matrix $[\; A \; \vdots \; I_n \;]$. Use row operations to change it to $\left[\; I_n \; \vdots \; A^{-1} \;\right]$. The row operations are effectively doing the following. Starting with $[\; A \; \vdots \; I_n \;]$, Steps 1 and 2 are like multiplying by $A^{-1}$ to yield

$$\left[\; A^{-1}A \; \vdots \; A^{-1}I_n \;\right] = \left[\; I_n \; \vdots \; A^{-1} \;\right].$$

We will demonstrate this with three examples.

| Example invert 1: Invert the matrix $\begin{bmatrix} 1 & -3 \\ -2 & 4 \end{bmatrix}$ | | | | |
|---|---|---|---|---|
| Form the initial matrices | 1 | -3 | 1 | 0 |
| | -2 | 4 | 0 | 1 |
| Multiply row 1 by (-2). Subtract the result from row 2. → | 1 | -3 | 1 | 0 |
| | 0 | -2 | 2 | 1 |
| Now multiply row 2 by 3/2. Subtract result from row 1 → | 1 | 0 | -2 | -1.5 |
| | 0 | -2 | 2 | 1 |
| Now divide row 2 by (-2) → | 1 | 0 | -2 | -1.5 |
| | 0 | 1 | -1 | -0.5 |

| | | | | | | |
|---|---|---|---|---|---|---|
| Example invert 2: Invert the matrix $\begin{bmatrix} 1 & 1/2 & 1/3 \\ 1/2 & 1/3 & 1/4 \\ 1/3 & 1/4 & 1/5 \end{bmatrix}$ | | | | | | |
| Form the initial matrix. | 1 | 1/2 | 1/3 | 1 | 0 | 0 |
| | 1/2 | 1/3 | 1/4 | 0 | 1 | 0 |
| | 1/3 | 1/4 | 1/5 | 0 | 0 | 1 |
| Step 1.1: Referring to the initial matrix, the goal is to force $a_{21} = 0$. The multiplier $= \dfrac{a_{21}}{a_{11}} = \dfrac{1}{2}$. Multiply row 1 by 1/2, and subtract it from row 2. | 1 | 1/2 | 1/3 | 1 | 0 | 0 |
| | 0 | 1/12 | 1/12 | -1/2 | 1 | 0 |
| | 1/3 | 1/4 | 1/5 | 0 | 0 | 1 |
| Step 1.2: Referring to the matrix of Step 1.1, the goal is to force $a_{31} = 0$. The multiplier $= \dfrac{a_{31}}{a_{11}} = \dfrac{1}{3}$. Multiply row 1 by 1/3 and subtract it from row 3. | 1 | 1/2 | 1/3 | 1 | 0 | 0 |
| | 0 | 1/12 | 1/12 | -1/2 | 1 | 0 |
| | 0 | 1/12 | 4/45 | -1/3 | 0 | 1 |
| Step 1.3: Referring to the matrix of Step 1.2, the goal is to force $a_{32} = 0$. The multiplier $= \dfrac{a_{32}}{a_{22}} = 1$. Multiply row 2 by 1 and subtract it from row 3. | 1 | 1/2 | 1/3 | 1 | 0 | 0 |
| | 0 | 1/12 | 1/12 | -1/2 | 1 | 0 |
| | 0 | 0 | 1/180 | 1/6 | -1 | 1 |
| Step 2.1: Referring to the matrix of Step 1.3, the goal is to force $a_{13} = 0$. The multiplier $= \dfrac{a_{13}}{a_{33}} = 60$. Multiply row 3 by 60 and subtract it from row 1. | 1 | 1/2 | 0 | -9 | 60 | -60 |
| | 0 | 1/12 | 1/12 | -1/2 | 1 | 0 |
| | 0 | 0 | 1/180 | 1/6 | -1 | 1 |
| Step 2.2: Referring to the matrix of Step 2.1, the goal is to force $a_{23} = 0$. The multiplier $= \dfrac{a_{23}}{a_{33}} = 15$. Multiply row 3 by 15 and subtract it from row 2. | 1 | 1/2 | 0 | -9 | 60 | -60 |
| | 0 | 1/12 | 0 | -3 | 16 | -15 |
| | 0 | 0 | 1/180 | 1/6 | -1 | 1 |
| Step 2.3: Referring to the matrix of Step 2.2, the goal is to force $a_{12} = 0$. The multiplier $= \dfrac{a_{12}}{a_{22}} = 6$. Multiply row 2 by 6 and subtract it from row 1. | 1 | 0 | 0 | 9 | -36 | 30 |
| | 0 | 1/12 | 0 | -3 | 16 | -15 |
| | 0 | 0 | 1/180 | 1/6 | -1 | 1 |
| Step 3: Referring to the matrix of Step 2.3, unitize the diagonal of the first three columns. | 1 | 0 | 0 | 9 | -36 | 30 |
| | 0 | 1 | 0 | -36 | 192 | -180 |
| | 0 | 0 | 1 | 30 | -180 | 180 |
| This is verified since $A^{-1}A = I_n$. Also see Example mult.5. | | | | | | |

| Example invert 3: Invert the matrix $\begin{bmatrix} 3 & -4 & -5 & 1 \\ 0 & 4 & 0 & 1 \\ 3 & 1 & 7 & 1 \\ 0 & 8 & 0 & 2 \end{bmatrix}.$ | | | | | | | | |
|---|---|---|---|---|---|---|---|---|

| | | | | | | | | |
|---|---|---|---|---|---|---|---|---|
| Form the initial matrix | 3 | -4 | -5 | 1 | 1 | 0 | 0 | 0 |
| | 0 | 4 | 0 | 1 | 0 | 1 | 0 | 0 |
| | 3 | 1 | 7 | 1 | 0 | 0 | 1 | 0 |
| | 0 | 8 | 0 | 2 | 0 | 0 | 0 | 1 |
| Step 1.1: Referring to the initial matrix, since $a_{21} = 0$, the goal is to force $a_{31} = 0$. The multiplier is $a_{31}/a_{11} = 1$. Multiply row 1 by 1, and subtract it from row 3. | 3 | -4 | -5 | 1 | 1 | 0 | 0 | 0 |
| | 0 | 4 | 0 | 1 | 0 | 1 | 0 | 0 |
| | 0 | 5 | 12 | 0 | -1 | 0 | 1 | 0 |
| | 0 | 8 | 0 | 2 | 0 | 0 | 0 | 1 |
| Step 1.2: Referring to the matrix of Step 1.1, since $a_{41} = 0$, the goal is to force $a_{32} = 0$. The multiplier is $a_{32}/a_{22} = 5/4$. Multiply row 2 by 5/4 and subtract it from row 3. | 3 | -4 | -5 | 1 | 1 | 0 | 0 | 0 |
| | 0 | 4 | 0 | 1 | 0 | 1 | 0 | 0 |
| | 0 | 0 | 12 | -5/4 | -1 | -5/4 | 1 | 0 |
| | 0 | 8 | 0 | 2 | 0 | 0 | 0 | 1 |
| Step 1.3: Referring to the matrix of Step 1.2, the goal is to force $a_{42} = 0$. The multiplier is $a_{42}/a_{22} = 2$. Multiply row 2 by 2 and subtract it from row 4. | 3 | -4 | -5 | 1 | 1 | 0 | 0 | 0 |
| | 0 | 4 | 0 | 1 | 0 | 1 | 0 | 0 |
| | 0 | 0 | 12 | -5/4 | -1 | -5/4 | 1 | 0 |
| | 0 | 0 | 0 | 0 | 0 | -2 | 0 | 1 |

Since $a_{43} = 0$, Step 2 can begin. But Step 2 cannot begin because the diagonal element $a_{44} = 0$. Therefore, the original matrix has no inverse.

## Determinants and Cramer's Rule

We started this chapter with a typical algebra problem:

$$v - 3w = 4$$
$$-2v + 4w = -6$$

We showed three ways to solve it: (1) substitution; (2) directly by row operations; and (3) by row operations for $A^{-1}$, followed by $x = A^{-1}b$.

In this section, we will show a fourth method. It's called Cramer's rule, and it uses determinants. To start the explanation, let's describe the above system in general.

$$a_{11}v + a_{12}w = k_1$$
$$a_{21}v + a_{22}w = k_2$$

Equation D.1

Using substitution, we can derive

$$v = \frac{(k_1 a_{22} - k_2 a_{12})}{(a_{11}a_{22} - a_{12}a_{21})} \quad \text{and} \quad w = \frac{(k_2 a_{11} - k_1 a_{21})}{(a_{11}a_{22} - a_{12}a_{21})}.$$

Equation D.2

We are ready to define a determinant. Suppose $A = \begin{bmatrix} a_{11} & a_{12} \\ a_{21} & a_{22} \end{bmatrix}$. By definition, the determinant of A is that number computed from the expression $(a_{11}a_{22} - a_{12}a_{21})$. Notice that this is the denominator of Equation D.2. That gives us a clue about solving equations using determinants. The determinant is given a symbol. The determinant of A is written $|A| = \begin{vmatrix} a_{11} & a_{12} \\ a_{21} & a_{22} \end{vmatrix} = (a_{11}a_{22} - a_{12}a_{21})$.

Using this symbol, Equation D.2 can be written as

$$v = \frac{(k_1 a_{22} - k_2 a_{12})}{\begin{vmatrix} a_{11} & a_{12} \\ a_{21} & a_{22} \end{vmatrix}} \quad \text{and} \quad w = \frac{(k_2 a_{11} - k_1 a_{21})}{\begin{vmatrix} a_{11} & a_{12} \\ a_{21} & a_{22} \end{vmatrix}}.$$

We turn our attention to the numerators of v and w. Let's multiply the first column of $|A|$ by v.

$$v * |A| = \begin{vmatrix} v * a_{11} & a_{12} \\ v * a_{21} & a_{22} \end{vmatrix} = v * (a_{11}a_{22} - a_{12}a_{21})$$

Let's now multiply the second column of $v * |A|$ by w, and add the result to the first column.

$$\begin{vmatrix} (v*a_{11}+w*a_{12}) & a_{12} \\ (v*a_{21}+w*a_{22}) & a_{22} \end{vmatrix} = (v*a_{11}+w*a_{12})*a_{22} - (v*a_{21}+w*a_{22})*a_{12}$$

$$= v*(a_{11}a_{22}-a_{12}a_{21}) + w*(a_{12}a_{22}-a_{12}a_{22})$$

$$= v*(a_{11}a_{22}-a_{12}a_{21}) = v*|A|$$

The value of $v*|A|$ did not change. From **Equation D.1**, $v*a_{11}+w*a_{12}=k_1$ and $v*a_{21}+w*a_{22}=k_2$.

Substituting these, $v*|A| = \begin{vmatrix} (v*a_{11}+w*a_{12}) & a_{12} \\ (v*a_{21}+w*a_{22}) & a_{22} \end{vmatrix} = \begin{vmatrix} k_1 & a_{12} \\ k_2 & a_{22} \end{vmatrix}$. Solving this for v yields

$$v = \frac{\begin{vmatrix} k_1 & a_{12} \\ k_2 & a_{22} \end{vmatrix}}{|A|} = \frac{\begin{vmatrix} k_1 & a_{12} \\ k_2 & a_{22} \end{vmatrix}}{\begin{vmatrix} a_{11} & a_{12} \\ a_{21} & a_{22} \end{vmatrix}} = \frac{(k_1 a_{22} - k_2 a_{12})}{(a_{11}a_{22} - a_{12}a_{21})}$$

This agrees with **Equation D.2**. If we do the same thing for w, we get

$$w = \frac{\begin{vmatrix} a_{11} & k_1 \\ a_{21} & k_2 \end{vmatrix}}{\begin{vmatrix} a_{11} & a_{12} \\ a_{21} & a_{22} \end{vmatrix}} = \frac{(k_2 a_{11} - k_1 a_{21})}{(a_{11}a_{22} - a_{12}a_{21})}$$

This also agrees with **Equation D.2**. The following table is a summary.

| Solving Second-Order Systems Using Cramer's Rule | |
|---|---|
| The Equations | Step 1: System Determinant |
| $a_{11}v + a_{12}w = k_1$ <br> $a_{21}v + a_{22}w = k_2$ | $|A| = \begin{vmatrix} a_{11} & a_{12} \\ a_{21} & a_{22} \end{vmatrix}$ |
| Step 2: First Variable (First Column) | Step 3: Second Variable (Second Column) |
| $v*|A| = \begin{vmatrix} k_1 & a_{12} \\ k_2 & a_{22} \end{vmatrix} \rightarrow v = \frac{\begin{vmatrix} k_1 & a_{12} \\ k_2 & a_{22} \end{vmatrix}}{\begin{vmatrix} a_{11} & a_{12} \\ a_{21} & a_{22} \end{vmatrix}}$ | $w*|A| = \begin{vmatrix} a_{11} & k_1 \\ a_{21} & k_2 \end{vmatrix} \rightarrow w = \frac{\begin{vmatrix} a_{11} & k_1 \\ a_{21} & k_2 \end{vmatrix}}{\begin{vmatrix} a_{11} & a_{12} \\ a_{21} & a_{22} \end{vmatrix}}$ |

Let's plug some numbers into this table.

| Solving Example lineq 1 Using Cramer's Rule | |
| --- | --- |
| The Equations<br><br>$v - 3w = 4$<br>$-2v + 4w = -6$ | Step 1: System Determinant<br><br>$\|A\| = \begin{vmatrix} 1 & -3 \\ -2 & 4 \end{vmatrix} = -2$ |
| Step 2: First Variable<br><br>$v*\|A\| = \begin{vmatrix} 4 & -3 \\ -6 & 4 \end{vmatrix} = -2 \;\rightarrow\; v = \dfrac{-2}{-2} = 1$ | Step 3: Second Variable<br><br>$w*\|A\| = \begin{vmatrix} 1 & 4 \\ -2 & -6 \end{vmatrix} = 2 \;\rightarrow\; w = \dfrac{2}{-2} = -1$ |
| This is the correct answer for Example lineq 1. | |

## The Determinant of a Third-Order System

Consider the third-order system.

$$a_{11}x_1 + a_{12}x_2 + a_{13}x_3 = k_1$$
$$a_{21}x_1 + a_{22}x_2 + a_{23}x_3 = k_2$$
$$a_{31}x_1 + a_{32}x_2 + a_{33}x_3 = k_3$$

The determinant for this system is

$$D = \begin{vmatrix} a_{11} & a_{12} & a_{13} \\ a_{21} & a_{22} & a_{23} \\ a_{31} & a_{32} & a_{33} \end{vmatrix}.$$

We will now show how to compute this determinant. Using the method of substitution, the denominator of the third-order system is

$$a_{11}a_{22}a_{33} + a_{12}a_{23}a_{31} + a_{13}a_{32}a_{21} - a_{13}a_{22}a_{31} - a_{12}a_{21}a_{33} - a_{11}a_{32}a_{23} \qquad \text{Equation D.3}$$

Let's group this expression.

$$a_{11}\left(a_{22}a_{33} - a_{32}a_{23}\right) - a_{21}\left(a_{12}a_{33} - a_{13}a_{32}\right) + a_{31}\left(a_{12}a_{23} - a_{13}a_{22}\right)$$

Using determinants, this expression can be written as

$$D = a_{11}\begin{vmatrix} a_{22} & a_{23} \\ a_{32} & a_{33} \end{vmatrix} - a_{21}\begin{vmatrix} a_{12} & a_{13} \\ a_{32} & a_{33} \end{vmatrix} + a_{31}\begin{vmatrix} a_{12} & a_{13} \\ a_{22} & a_{23} \end{vmatrix}.$$

Introducing some notation, $\qquad D = a_{11}D_{11} + \left(-a_{21}D_{21}\right) + a_{31}D_{31}.$ $\qquad$ Equation D.4

Equation D.4 has three terms. They are called minors. Each minor has this form:

$$(-1)^{i+j} * a_{ij} * D_{ij}$$

where $D_{ij}$ is the determinant that remains after row $i$ and column $j$ of D are crossed out. For example,

$$D = \begin{vmatrix} a_{11} & a_{12} & a_{13} \\ a_{21} & a_{22} & a_{23} \\ a_{31} & a_{32} & a_{33} \end{vmatrix}, \text{ then } D_{11} = \begin{vmatrix} \cancel{a_{11}} & \cancel{a_{12}} & \cancel{a_{13}} \\ \cancel{a_{21}} & a_{22} & a_{23} \\ \cancel{a_{31}} & a_{32} & a_{33} \end{vmatrix} = \begin{vmatrix} a_{22} & a_{23} \\ a_{32} & a_{33} \end{vmatrix}.$$

In Equation D.4, the elements ($a_{11}$, $a_{21}$ and $a_{31}$) are all from the first column of D. Hence, D is said to be expanded about its first column.

Let's repeat Equation D.3.

$$a_{11}a_{22}a_{33} + a_{12}a_{23}a_{31} + a_{13}a_{32}a_{21} - a_{13}a_{22}a_{31} - a_{12}a_{21}a_{33} - a_{11}a_{32}a_{23} \qquad \text{Equation D.3}$$

Let's group it another way.

$$a_{11}\left(a_{22}a_{33} - a_{32}a_{23}\right) - a_{12}\left(a_{21}a_{33} - a_{23}a_{31}\right) + a_{13}\left(a_{21}a_{32} - a_{22}a_{31}\right)$$

Using determinants,
$$D = a_{11}\begin{vmatrix} a_{22} & a_{23} \\ a_{32} & a_{33} \end{vmatrix} - a_{12}\begin{vmatrix} a_{21} & a_{23} \\ a_{31} & a_{33} \end{vmatrix} + a_{13}\begin{vmatrix} a_{21} & a_{22} \\ a_{31} & a_{32} \end{vmatrix}$$

$$D = a_{11}D_{11} + \left(-a_{12}D_{12}\right) + a_{13}D_{13} \qquad \text{Equation D.5}$$

The Equation D.5 elements ($a_{11}$, $a_{12}$ and $a_{13}$) are all from the first row of D. Hence, D is said to be expanded about its first row. The value of D can be computed from either Equation D.4 or D.5. Since other rows or columns can be used, there are many ways to compute the third-order determinant.

We have shown that the value of a third-order determinant can be computed by expanding it in minors. The following is a summary of this process. The determinant is

$$D = \begin{vmatrix} a_{11} & a_{12} & a_{13} \\ a_{21} & a_{22} & a_{23} \\ a_{31} & a_{32} & a_{33} \end{vmatrix}$$

Equation D.4 applies when D is expanded about its first column as shown in the following table.

| Equation D.4: $D = a_{11}D_{11} + \left(-a_{21}D_{21}\right) + a_{31}D_{31}$ | | |
|---|---|---|
| $D_{11} = \begin{vmatrix} \cancel{a_{11}} & \cancel{a_{12}} & \cancel{a_{13}} \\ \cancel{a_{21}} & a_{22} & a_{23} \\ \cancel{a_{31}} & a_{32} & a_{33} \end{vmatrix} = \begin{vmatrix} a_{22} & a_{23} \\ a_{32} & a_{33} \end{vmatrix}$ | $D_{21} = \begin{vmatrix} \cancel{a_{11}} & a_{12} & a_{13} \\ \cancel{a_{21}} & \cancel{a_{22}} & \cancel{a_{23}} \\ \cancel{a_{31}} & a_{32} & a_{33} \end{vmatrix} = \begin{vmatrix} a_{12} & a_{13} \\ a_{32} & a_{33} \end{vmatrix}$ | $D_{31} = \begin{vmatrix} \cancel{a_{11}} & a_{12} & a_{13} \\ \cancel{a_{21}} & a_{22} & a_{23} \\ \cancel{a_{31}} & \cancel{a_{32}} & \cancel{a_{33}} \end{vmatrix} = \begin{vmatrix} a_{12} & a_{13} \\ a_{22} & a_{23} \end{vmatrix}$ |
| $D = a_{11}\begin{vmatrix} a_{22} & a_{23} \\ a_{32} & a_{33} \end{vmatrix} - a_{21}\begin{vmatrix} a_{12} & a_{13} \\ a_{32} & a_{33} \end{vmatrix} + a_{31}\begin{vmatrix} a_{12} & a_{13} \\ a_{22} & a_{23} \end{vmatrix}$ | | |

Let's plug in some numbers. Expand $D = \begin{vmatrix} 1 & 0 & 1 \\ 2 & 3 & 1 \\ 3 & 5 & 2 \end{vmatrix}$ about its first column. This is shown as follows.

| From Equation D.4: $D = 1*D_{11} - 2*D_{21} + 3*D_{31}$ | | |
|---|---|---|
| $D_{11} = \begin{vmatrix} \cancel{1} & \cancel{0} & \cancel{1} \\ \cancel{2} & 3 & 1 \\ \cancel{3} & 5 & 2 \end{vmatrix} = \begin{vmatrix} 3 & 1 \\ 5 & 2 \end{vmatrix}$ | $D_{21} = \begin{vmatrix} \cancel{1} & 0 & 1 \\ \cancel{2} & \cancel{3} & \cancel{1} \\ \cancel{3} & 5 & 2 \end{vmatrix} = \begin{vmatrix} 0 & 1 \\ 5 & 2 \end{vmatrix}$ | $D_{31} = \begin{vmatrix} \cancel{1} & 0 & 1 \\ \cancel{2} & 3 & 1 \\ \cancel{3} & \cancel{5} & \cancel{2} \end{vmatrix} = \begin{vmatrix} 0 & 1 \\ 3 & 1 \end{vmatrix}$ |

$$D = 1*\begin{vmatrix} 3 & 1 \\ 5 & 2 \end{vmatrix} - 2*\begin{vmatrix} 0 & 1 \\ 5 & 2 \end{vmatrix} + 3*\begin{vmatrix} 0 & 1 \\ 3 & 1 \end{vmatrix} = 1*(6-5) - 2*(0-5) + 3*(0-3) = 2$$

Let's expand the same determinant about its first row. This is shown as follows.

| From Equation D.5: $D = 1*D_{11} - 0*D_{12} + 1*D_{13}$ | | |
|---|---|---|
| $D_{11} = \begin{vmatrix} \cancel{1} & \cancel{0} & \cancel{1} \\ \cancel{2} & 3 & 1 \\ \cancel{3} & 5 & 2 \end{vmatrix} = \begin{vmatrix} 3 & 1 \\ 5 & 2 \end{vmatrix}$ | $D_{12} = \begin{vmatrix} \cancel{1} & \cancel{0} & \cancel{1} \\ 2 & \cancel{3} & 1 \\ 3 & \cancel{5} & 2 \end{vmatrix} = \begin{vmatrix} 2 & 1 \\ 3 & 2 \end{vmatrix}$ | $D_{13} = \begin{vmatrix} \cancel{1} & \cancel{0} & \cancel{1} \\ 2 & 3 & \cancel{1} \\ 3 & 5 & \cancel{2} \end{vmatrix} = \begin{vmatrix} 2 & 3 \\ 3 & 5 \end{vmatrix}$ |

$$D = 1*\begin{vmatrix} 3 & 1 \\ 5 & 2 \end{vmatrix} - 0*\begin{vmatrix} 2 & 1 \\ 3 & 2 \end{vmatrix} + 1*\begin{vmatrix} 2 & 3 \\ 3 & 5 \end{vmatrix} = 1*(6-5) + 1*(10-9) = 2$$

Either way, the value of D is the same.

## Solving Third-Order Systems Using Cramer's Rule

Applying Cramer's rule to third-order systems follows the pattern set for second-order systems.

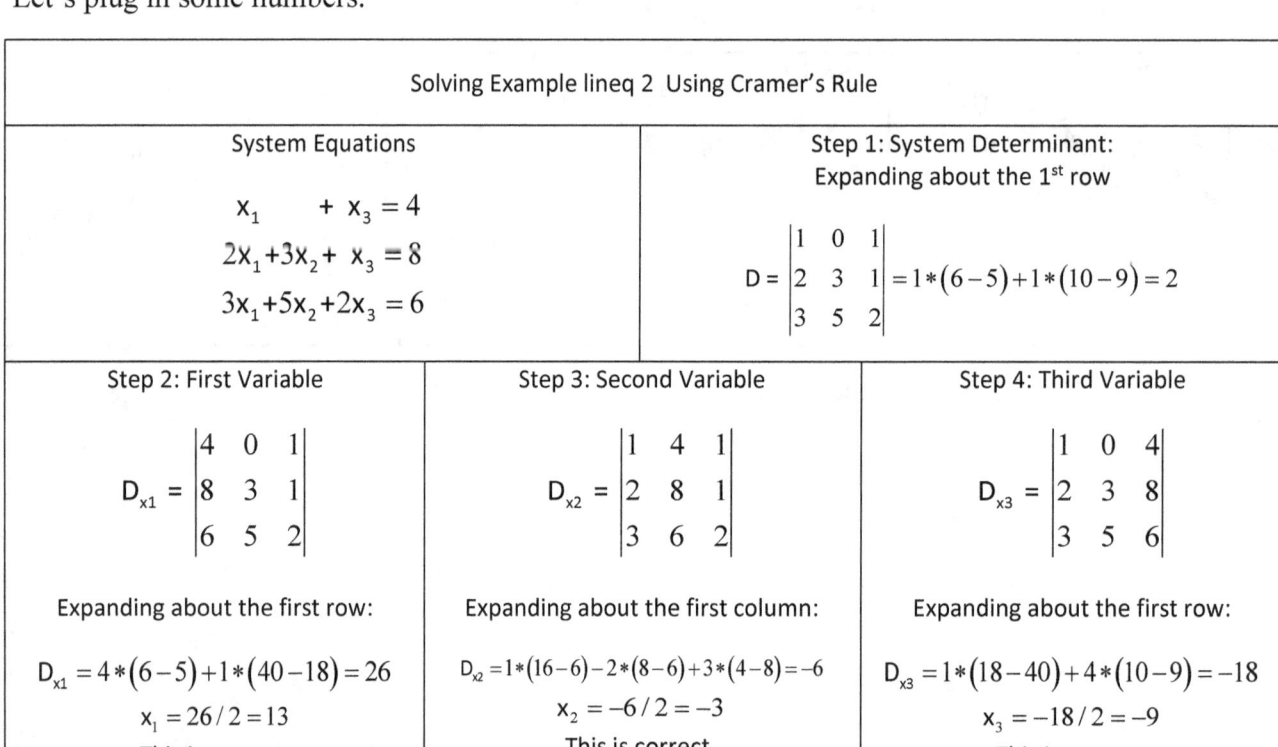

Let's plug in some numbers.

## Solving Third-Order Systems Using Cramer's Rule

Applying Cramer's rule to third-order systems follows the pattern set for second-order systems.

| Solving Third-Order Systems Using Cramer's Rule | |
|---|---|
| **System Equations** | **Step 1: System Determinant** |
| $a_{11}x_1 + a_{12}x_2 + a_{13}x_3 = k_1$ <br> $a_{21}x_1 + a_{22}x_2 + a_{23}x_3 = k_2$ <br> $a_{31}x_1 + a_{32}x_2 + a_{33}x_3 = k_3$ | $D = \begin{vmatrix} a_{11} & a_{12} & a_{13} \\ a_{21} & a_{22} & a_{23} \\ a_{31} & a_{32} & a_{33} \end{vmatrix}$ |

| **Step 2: First Variable (1st Column)** | **Step 3: Second Variable (2nd Column)** | **Step 4: Third Variable (3rd Column)** |
|---|---|---|
| $D_{x1} = \begin{vmatrix} k_1 & a_{12} & a_{13} \\ k_2 & a_{22} & a_{23} \\ k_3 & a_{32} & a_{33} \end{vmatrix} \rightarrow x_1 = \dfrac{D_{x1}}{D}$ | $D_{x2} = \begin{vmatrix} a_{11} & k_1 & a_{13} \\ a_{21} & k_2 & a_{23} \\ a_{31} & k_3 & a_{33} \end{vmatrix} \rightarrow x_2 = \dfrac{D_{x2}}{D}$ | $D_{x3} = \begin{vmatrix} a_{11} & a_{12} & k_1 \\ a_{21} & a_{22} & k_2 \\ a_{31} & a_{32} & k_3 \end{vmatrix} \rightarrow x_3 = \dfrac{D_{x3}}{D}$ |

Let's plug in some numbers.

| Solving Example lineq 2 Using Cramer's Rule | |
|---|---|
| **System Equations** | **Step 1: System Determinant:** <br> **Expanding about the 1st row** |
| $x_1 \quad\quad + x_3 = 4$ <br> $2x_1 + 3x_2 + x_3 = 8$ <br> $3x_1 + 5x_2 + 2x_3 = 6$ | $D = \begin{vmatrix} 1 & 0 & 1 \\ 2 & 3 & 1 \\ 3 & 5 & 2 \end{vmatrix} = 1*(6-5) + 1*(10-9) = 2$ |

| **Step 2: First Variable** | **Step 3: Second Variable** | **Step 4: Third Variable** |
|---|---|---|
| $D_{x1} = \begin{vmatrix} 4 & 0 & 1 \\ 8 & 3 & 1 \\ 6 & 5 & 2 \end{vmatrix}$ | $D_{x2} = \begin{vmatrix} 1 & 4 & 1 \\ 2 & 8 & 1 \\ 3 & 6 & 2 \end{vmatrix}$ | $D_{x3} = \begin{vmatrix} 1 & 0 & 4 \\ 2 & 3 & 8 \\ 3 & 5 & 6 \end{vmatrix}$ |
| Expanding about the first row: | Expanding about the first column: | Expanding about the first row: |
| $D_{x1} = 4*(6-5) + 1*(40-18) = 26$ <br> $x_1 = 26/2 = 13$ <br> This is correct. | $D_{x2} = 1*(16-6) - 2*(8-6) + 3*(4-8) = -6$ <br> $x_2 = -6/2 = -3$ <br> This is correct. | $D_{x3} = 1*(18-40) + 4*(10-9) = -18$ <br> $x_3 = -18/2 = -9$ <br> This is correct. |

## How to Evaluate Higher-Order Determinants

Let's evaluate a fourth-order determinant by expanding it about its fourth column.

$$D = \begin{vmatrix} c_{11} & c_{12} & c_{13} & c_{14} \\ c_{21} & c_{22} & c_{23} & c_{24} \\ c_{31} & c_{32} & c_{33} & c_{34} \\ c_{41} & c_{42} & c_{43} & c_{44} \end{vmatrix} = -c_{14}D_{14} + c_{24}D_{24} - c_{34}D_{34} + c_{44}D_{44}$$

where $D_{14} = \begin{vmatrix} c_{21} & c_{22} & c_{23} \\ c_{31} & c_{32} & c_{33} \\ c_{41} & c_{42} & c_{43} \end{vmatrix}$, $D_{24} = \begin{vmatrix} c_{11} & c_{12} & c_{13} \\ c_{31} & c_{32} & c_{33} \\ c_{41} & c_{42} & c_{43} \end{vmatrix}$, $D_{34} = \begin{vmatrix} c_{11} & c_{12} & c_{13} \\ c_{21} & c_{22} & c_{23} \\ c_{41} & c_{42} & c_{43} \end{vmatrix}$ and $D_{44} = \begin{vmatrix} c_{11} & c_{12} & c_{13} \\ c_{21} & c_{22} & c_{23} \\ c_{31} & c_{32} & c_{33} \end{vmatrix}$.

Each of these third-order determinants is then expanded about one of its rows or columns. This will yield second-order determinants that are easily evaluated. Determinants of all sizes are evaluated in this manner.

## Solving $n^{th}$-Order Equations Using Cramer's Rule

$$\begin{aligned} a_{11}x_1 + a_{12}x_2 + \bullet\bullet\bullet + a_{1n}x_n &= k_1 \\ a_{21}x_1 + a_{22}x_2 + \bullet\bullet\bullet + a_{2n}x_n &= k_2 \\ &\vdots \\ a_{n1}x_1 + a_{n2}x_2 + \bullet\bullet\bullet + a_{nn}x_n &= k_n \end{aligned}$$

$$D = \begin{vmatrix} a_{11} & a_{12} & \bullet & \bullet & \bullet & a_{1n} \\ a_{21} & a_{22} & \bullet & \bullet & \bullet & a_{2n} \\ \bullet & \bullet & \bullet & \bullet & \bullet & \bullet \\ \bullet & \bullet & \bullet & \bullet & \bullet & \bullet \\ \bullet & \bullet & \bullet & \bullet & \bullet & \bullet \\ a_{n1} & a_{n2} & \bullet & \bullet & \bullet & a_{nn} \end{vmatrix}$$

$$D_{X1} = \begin{vmatrix} k_1 & a_{12} & \bullet & \bullet & \bullet & a_{1n} \\ k_2 & a_{22} & \bullet & \bullet & \bullet & a_{2n} \\ \bullet & \bullet & \bullet & \bullet & \bullet & \bullet \\ \bullet & \bullet & \bullet & \bullet & \bullet & \bullet \\ \bullet & \bullet & \bullet & \bullet & \bullet & \bullet \\ k_n & a_{n2} & \bullet & \bullet & \bullet & a_{nn} \end{vmatrix} \Rightarrow X_1 = \frac{D_{X1}}{D}$$

$$\vdots$$

$$D_{Xn} = \begin{vmatrix} a_{11} & a_{12} & \bullet & \bullet & \bullet & k_1 \\ a_{21} & a_{22} & \bullet & \bullet & \bullet & k_2 \\ \bullet & \bullet & \bullet & \bullet & \bullet & \bullet \\ \bullet & \bullet & \bullet & \bullet & \bullet & \bullet \\ \bullet & \bullet & \bullet & \bullet & \bullet & \bullet \\ a_{n1} & a_{n2} & \bullet & \bullet & \bullet & k_n \end{vmatrix} \Rightarrow X_n = \frac{D_{Xn}}{D}$$

## How to Evaluate a Determinant That Has All Zeros below Its Diagonal

This situation will be encountered in chapter 5 where we will find roots of polynomials. Consider the following fourth-order determinant that has been expanded about its first column.

$$D = \begin{vmatrix} c_{11} & c_{12} & c_{13} & c_{14} \\ 0 & c_{22} & c_{23} & c_{24} \\ 0 & 0 & c_{33} & c_{34} \\ 0 & 0 & 0 & c_{44} \end{vmatrix} = c_{11} * \begin{vmatrix} c_{22} & c_{23} & c_{24} \\ 0 & c_{33} & c_{34} \\ 0 & 0 & c_{44} \end{vmatrix}.$$

If we continue this process, we get

$$D = c_{11} * \begin{vmatrix} c_{22} & c_{23} & c_{24} \\ 0 & c_{33} & c_{34} \\ 0 & 0 & c_{44} \end{vmatrix} = c_{11} * c_{22} * \begin{vmatrix} c_{33} & c_{34} \\ 0 & c_{44} \end{vmatrix} = c_{11} * c_{22} * c_{33} * c_{44}.$$

The conclusion is that the value of a determinant having zeros below its diagonal, is the product of the numbers on its diagonal.

## When the Determinant Is Zero

When the determinant is zero, Cramer's rule can't be used. One of two situations exists. These are the focus of this section.

**Situation 1:** In **Example lineq 4**, no solution could be found. The equations to be solved are

$$\begin{bmatrix} 4 & 2 & -4 \\ 2 & 1 & -1 \\ 6 & 3 & -5 \end{bmatrix} \begin{bmatrix} u \\ v \\ w \end{bmatrix} = \begin{bmatrix} 6 \\ 4 \\ 6 \end{bmatrix}$$

The determinant is $\begin{vmatrix} 4 & 2 & -4 \\ 2 & 1 & -1 \\ 6 & 3 & -5 \end{vmatrix} = (-20 - 12 - 24) - (-24 - 20 - 12) = 0$.

In equation form,

$$4u + 2v - 4w = 6 \qquad \text{(Equation } a\text{)}$$
$$2u + v - w = 4 \qquad \text{(Equation } b\text{)}$$
$$6u + 3v - 5w = 6 \qquad \text{(Equation } c\text{)}$$

Let's solve these using the method of substitution.

- From **Equation b:** $v = 4 - 2u + w$.
- Substituting v into **Equation a:** $4u + 2(4 - 2u + w) - 4w = 6$. This yields w = 1.
- Substituting v into **Equation c:** $6u + 3(4 - 2u + w) - 5w = 6$. This yields w = 3.

These answers are contradictory. The equations are inconsistent and no solution exists.

Situation 2: The equations to be solved are

$$\begin{bmatrix} 1 & 0 & 1 \\ 5 & 10 & 3 \\ 3 & 5 & 2 \end{bmatrix} \begin{bmatrix} u \\ v \\ w \end{bmatrix} = \begin{bmatrix} 4 \\ 8 \\ 6 \end{bmatrix}$$

The determinant is
$$\begin{vmatrix} 1 & 0 & 1 \\ 5 & 10 & 3 \\ 3 & 5 & 2 \end{vmatrix} = (20+25)-(30+15) = 0.$$

In equation form,

$$u + 0v + w = 4 \qquad \text{(Equation d)}$$
$$5u + 10v + 3w = 8 \qquad \text{(Equation e)}$$
$$3u + 5v + 2w = 6 \qquad \text{(Equation f)}$$

Let's solve these using the method of substitution.

- From Equation d: $u = 4 - w$.
- Substituting u into Equation e, $5(4-w)+10v+3w = 8$. This yields $10v - 2w = -12$.
- Substituting u into Equation f, $3(4-w)+ 5v+2w = 6$. This yields $5v - w = -6$.

The equation $(10v - 2w = -12)$ is two times the equation $(5v - w = -6)$. They are not independent. Hence there is not enough information to solve them. So, we can add our own information. Pick w = 1. This would yield u = 3 and v = −1. This is a legitimate solution. Obviously, there is an infinite number of solutions to this underdetermined case.

## Conclusions about Determinants

- When the determinant is zero, there is either no solution, or an infinite number of solutions.
- When the determinant is not equal to zero, there is only one solution.

## Homogeneous Equations

The following is a homogeneous matrix equation. $[A][x] = [0]$. $x = 0$ is always a solution. What about the other solutions? We will use examples to answer this equation.

Example H.1: The determinant of A is not equal to zero.

We are given 
$$\begin{bmatrix} 1 & 0 & 1 \\ 2 & 3 & 1 \\ 3 & 5 & 2 \end{bmatrix} \begin{bmatrix} x_1 \\ x_2 \\ x_3 \end{bmatrix} = \begin{bmatrix} 0 \\ 0 \\ 0 \end{bmatrix}$$

$$|A| = \begin{vmatrix} 1 & 0 & 1 \\ 2 & 3 & 1 \\ 3 & 5 & 2 \end{vmatrix} = 2$$

We will first apply Cramer's rule.

| | | |
|---|---|---|
| $x_1 = \dfrac{\begin{vmatrix} 0 & 0 & 1 \\ 0 & 3 & 1 \\ 0 & 5 & 2 \end{vmatrix}}{|A|} = \dfrac{0}{2} = 0$ | $x_2 = \dfrac{\begin{vmatrix} 1 & 0 & 1 \\ 2 & 0 & 1 \\ 3 & 0 & 2 \end{vmatrix}}{|A|} = \dfrac{0}{2} = 0$ | $x_3 = \dfrac{\begin{vmatrix} 1 & 0 & 0 \\ 2 & 3 & 0 \\ 3 & 5 & 0 \end{vmatrix}}{|A|} = \dfrac{0}{2} = 0$ |

We will now use the substitution method.

- From the first equation, $x_1 = -x_3$.

- Substituting into the second equation, $2(-x_3) + 3x_2 + x_3 = 0 \rightarrow 3x_2 - x_3 = 0$      **Equation one**

- Substituting into the third equation, $3(-x_3) + 5x_2 + 2x_3 = 0 \rightarrow 5x_2 - x_3 = 0$      **Equation two**

**Equations one** and **two** are contradictory. Hence, no non-zero solution is possible. This example shows that when $|A| \neq 0$, the only solution to a homogeneous equation is the zero solution.

Example H.2: The determinant of A is equal to zero.

We are given 
$$\begin{bmatrix} 1 & 0 & 1 \\ 5 & 10 & 3 \\ 3 & 5 & 2 \end{bmatrix} \begin{bmatrix} x_1 \\ x_2 \\ x_3 \end{bmatrix} = \begin{bmatrix} 0 \\ 0 \\ 0 \end{bmatrix}$$

$$|A| = \begin{vmatrix} 1 & 0 & 1 \\ 5 & 10 & 3 \\ 3 & 5 & 2 \end{vmatrix} = 0$$

Cramer's rule can't be used. Let's use the substitution method.

- From the first equation, $x_1 = -x_3$.

- Substituting into the second equation, $5(-x_3) + 10x_2 + 3x_3 = 0 \rightarrow 10x_2 - 2x_3 = 0$     Equation three

- Substituting into the third equation,     $3(-x_3) + 5x_2 + 2x_3 = 0 \rightarrow 5x_2 - x_3 = 0$     Equation four

Equations three and four are the same. We have boiled the original equations down to

$$x_3 = 5x_2 \quad \text{and} \quad x_1 = -x_3.$$

These are two equations with three unknowns. We don't have enough information to solve them. Hence, we can add our own information. We can pick $x_2 = 1$. This would make $x_3 = 5$ and $x_1 = -5$. This is one valid solution to the original homogeneous equations. Obviously there are many more. This case shows that when $|A| = 0$, there are other solutions besides the zero solution.

Here is a summary for homogeneous equations:

- The zero solution always exists.
- If $|A| = 0$, there is either no solution, or an infinite number of solutions.
- If $|A| \neq 0$, the only solution is the zero solution.

## Summary

We will summarize this chapter using this problem. Solve these equations.

$$v - 3w = 4$$
$$-2v + 4w = -6$$

Method 1: Solve them using the substitution method.

Step 1: Solve the first equation for v.          $v = 4 + 3w$.
Step 2: Substitute this into the second equation.   $-2*(4+3w)+4w = -6$.
Step 3: Solve the Step 2 equation for w.          $w = -1$.
Step 4: Substitute this into the Step 1 equation.   $v = 1$.

Method 2: Solve them using the basic row operations.

| The initial matrices $[A \mid b]$ | 1 | -3 | 4 |
|---|---|---|---|
|  | -2 | 4 | -6 |
| The final matrices $[I \mid A^{-1}b = x]$ | 1 | 0 | 1 |
|  | 0 | 1 | -1 |

Method 3: Solve them using the basic row operations by first computing $A^{-1}$.

| The initial matrices $[A \mid I]$ | 1 | -3 | 1 | 0 |
|---|---|---|---|---|
|  | -2 | 4 | 0 | 1 |
| The final matrices $[I \mid A^{-1}]$ | 1 | 0 | -2 | -1.5 |
|  | 0 | 1 | -1 | -0.5 |

$$\text{Since } x = A^{-1}b, \text{ then } x = \begin{bmatrix} -2 & -1.5 \\ -1 & -0.5 \end{bmatrix}\begin{bmatrix} 4 \\ -6 \end{bmatrix} = \begin{bmatrix} 1 \\ -1 \end{bmatrix}$$

Method 4: Solve them using Cramer's rule.

| System Equations | Step 1: System Determinant |
|---|---|
| $v - 3w = 4$ <br> $-2v + 4w = -6$ | $|A| = \begin{vmatrix} 1 & -3 \\ -2 & 4 \end{vmatrix} = -2$ |
| **Step 2: First Variable** | **Step 3: Second Variable** |
| $v*|A| = \begin{vmatrix} 4 & -3 \\ -6 & 4 \end{vmatrix} = -2 \ \rightarrow \ v = \dfrac{-2}{-2} = 1$ | $w*|A| = \begin{vmatrix} 1 & 4 \\ -2 & -6 \end{vmatrix} = 2 \ \rightarrow \ w = \dfrac{2}{-2} = -1$ |

Final Note: Computing the determinant is a quick way to find out if there is a solution.

# Equations and Matrices – Programming in Chapter 3

We will now present the three programs that were used to do the math in this chapter.

- A program to multiply matrices.
- A program to solve the equation Ax = b.
- A program to invert matrices.

These programs introduce the following VBA syntax.

- The Call Statement.
- The Dim Statement.
- The If-Then-Else Statement.
- Non-numeric data.

## The Call Statement

So far, all of the programs have been single programs as shown in the following table.

```
Sub main()

    { VBA statements}

End Sub
```

We will now show how a program can be modularized. A main program calls a subprogram to do a task. After the task, control returns to the main. The following table illustrates this.

| | | |
|---|---|---|
| Start | Sub AAA() | |
| Step 1 | a2 = 4 | |
| Step 2 | Call BBB(b2, a2) | At Step 2, the program transfers to the subprogram called BBB. Enclosed within parentheses are the variables a2 and b2. In Sub BBB, a2 is printed and b2 is set to a value. Then the program returns to Sub AAA where b2 is printed and the program ends. |
| Step 7 | Cells(2, 1) = b2 | |
| Step 8 | End Sub | |
| | | |
| Step 3 | Sub BBB(b2, a2) | |
| Step 4 | Cells(1, 1) = a2 | |
| Step 5 | b2 = 5 | |
| Step 6 | End Sub | |
| Notes | | |
| • The argument list contains all of the variables that are transferred to and from.<br>• There is no limit on the number of Call statements or subprograms.<br>• Subprograms can call other subprograms.<br>• Except for the argument list, values of the variables in a subprogram must be reset or recomputed each time the subprogram is entered.<br>• The argument list may be empty.<br>• The calling program and the called program should be in the same code window.<br>• Only the programs with empty parentheses are listed in the Macros window. | | |

The following table contains a listing of an easy-to-follow program. It starts with a main program called zero. There are two subprograms called one and two. These three parts pass around a variable named A1. Starting with A1 = 0, each program increases the value of A1 until the final value is printed out in zero and the whole program ends. Read the code and satisfy yourself that the final value is A1 = 9.

| Program Call_1 | |
|---|---|
| Sub zero()<br>  A1 = 0<br>  Call one(A1)<br>  Cells(1, 1) = A1<br>End Sub<br><br>Sub one(A1)<br>  A1 = A1 + 3<br>  Call two(A1)<br>  A1 = A1 + 3<br>End Sub<br><br>Sub two(A1)<br>  A1 = A1 + 3<br>End Sub | Note that the three parts should be in the same code window. |

The following program produces identical results to the above program. It shows that the variable A1 is really an address in memory. In **Sub one**, this address is called B1. In **Sub two**, it is called C1.

| Program Call_2 |
|---|
| Sub zero()<br>  A1 = 0<br>  Call one(A1)<br>  Cells(1, 1) = A1<br>End Sub<br><br>Sub one(B1)<br>  B1 = B1 + 3<br>  Call two(B1)<br>  B1 = B1 + 3<br>End Sub<br><br>Sub two(C1)<br>  C1 = C1 + 3<br>End Sub |

## The Dim Statement

A matrix is a rectangular array of numbers. Suppose we have a row matrix with four numbers:

$$aa = \begin{bmatrix} 1 & 3 & 5 & 7 \end{bmatrix}$$

Before we can use **aa** in a program, we have to declare how many memory locations it will occupy. We do this using the **Dim** statement. The following code fragment shows this.

| | |
|---|---|
| Dim aa(4)<br>aa(1) = 1: aa(2) = 3: aa(3) = 5: aa(4) = 7<br>b = 2 * aa(3) | •The colon is a separator that allows more than one statement per line.<br>• The value of b = 10. |

Suppose we have a 2x3 matrix: $bb = \begin{bmatrix} 1 & 3 & 5 \\ 2 & 4 & 6 \end{bmatrix}$. The following is a code fragment that uses **bb** and a 4x1 matrix called **cc**.

| | |
|---|---|
| Dim bb(2, 3), cc(4, 1)<br>bb(1, 1) = 1: bb(1, 2) = 3: bb(1, 3) = 5<br>bb(2, 1) = 2: bb(2, 2) = 4: bb(2, 3) = 6<br>cc(2, 1) = bb(1, 1) + bb(1, 2) | • The value of cc(2, 1) = 4.<br><br>• The Dim statement may specify more locations than are actually needed. |

Program Call_2 uses the single variable A1. The following is a similar program that declares A1 to be a matrix that can contain fifty numbers. This program uses only the fourth entry of A1.

| Program Call_3 | |
|---|---|
| Sub zero()<br>Dim A1(50)<br>  xx = Array(-8, 3, 1)<br>  A1(4) = 0<br>  Call one(A1)<br>  Cells(1, 1) = A1(4)<br>  Cells(2, 1) = xx(0)<br>End Sub<br>Sub one(B1)<br>  B1(4) = B1(4) + 3<br>  Call two(B1)<br>  B1(4) = B1(4) + 3<br>End Sub<br>Sub two(C1)<br>Dim D1(10)<br>  D1(5) = 3<br>  C1(4) = C1(4) + D1(5)<br>End Sub | • Before a matrix can be used, it must be declared in a Dim statement. This declaration appears only in the program (or subprogram) wherein the matrix first appears.<br><br>• Matrix D1 that is declared in Sub two, is only valid in Sub two.<br><br>• The final value for A1(4) is 9.<br><br>• Note that xx(0) is printed out. Its value = −8. Though it acts like a matrix, xx comes from an Array function. The thing to know is that the output of an Array function does not appear in a Dim statement.<br><br>• See section 8.3 for more syntax on the Dim statement. |

## The If-Then-Else Statement

This VBA statement allows a statement or block of statements to be executed based on the true or false value of an expression. The following table shows a few examples.

| | |
|---|---|
| If xx < 10 Then<br>    yy = 1<br>    zz = 1<br>End If | If xx is less than 10, the enclosed block of statements will be executed. If xx is not, the program branches to the statement after End If. |
| If xx <> 0 Then yy = 1 | If xx is not equal to zero, then yy = 1. This statement has no End If and allows only one statement to be executed. |
| If xx = 0 Then<br>    yy = 1<br>    zz = 1<br>Else<br>    ww = 2<br>    zz = 2<br>End If | This statement allows a choice of statements to be executed. |
| See section 8.6 for more syntax on If-Then-Else. ||

## Non-numeric Data

This type of data is commonly used for titles and identifiers. It is enclosed within quotation marks. As an example, the statement $Cells(1,1) = $ "variable X" writes the identifier in cell (1,1).

## A Program to Multiply Matrices

We can now present a program that multiplies two matrices. The following table contains a code fragment. It uses three nested **For-Next** loops. It again shows that the *equal sign* means *is replaced by*.

<table>
<tr>
<td>

```
For i = 1 To M
  For j = 1 To N
    AB(i, j) = 0
    For k = 1 To L
      AB(i, j) = AB(i, j) + A(i, k) * B(k, j)
    Next k
  Next j
Next i
```

</td>
<td>

$$[AB] = [A] \ [B]$$
$$M \times N \quad M \times L \quad L \times N$$

• This program shows how to implement the summation operation. For each *i* and *j*,

$$AB(i,j) = \sum_{k=1}^{L} A(i,k) * B(k,j)$$

</td>
</tr>
<tr>
<td colspan="2">

• The number of columns of A must equal the number of rows of B.
• The number of rows of AB equals the number of rows of A.
• The number of columns of AB equals the number of columns of B.

</td>
</tr>
</table>

The following table contains the listing of the total program.

| Code, Page 1 of 2 | Code, Page 2 of 2 |
|---|---|
| <pre>Sub mult_main()<br> Dim A(20, 21), B(21, 22), AB(20, 22)<br> M = Cells(1, 1): L = Cells(1, 2): N = Cells(1, 3)<br> For i = 1 To M<br>   For k = 1 To L<br>     A(i, k) = Cells(i + 1, k)          ' read A<br>   Next k<br> Next i<br> For k = 1 To L<br>   For j = 1 To N<br>     B(k, j) = Cells(k + 1, j + L + 1)    ' read B<br>   Next j<br> Next k<br> Call mult(A, B, AB, M, L, N)<br> For i = 1 To M<br>   For j = 1 To N<br>     Cells(i + M + 2, j) = AB(i, j)<br>   Next j<br> Next i<br>End Sub</pre> | <pre>Sub mult(A, B, AB, M, L, N)<br> For i = 1 To M<br>   For j = 1 To N<br>     AB(i, j) = 0<br>     For k = 1 To L<br>       AB(i, j) = AB(i, j) + A(i, k) * B(k, j)<br>     Next k<br>   Next j<br> Next i<br>End Sub</pre> |
| Note: An apostrophe causes the rest of the line to be ignored. Use this to add comments to the code. ||

We will use this program to multiply the following two matrices.

$$A = \begin{bmatrix} 1 & 1/2 & 1/3 \\ 1/2 & 1/3 & 1/4 \\ 1/3 & 1/4 & 1/5 \end{bmatrix} \quad \text{and} \quad B = \begin{bmatrix} 9 & -36 & 30 \\ -36 & 192 & -180 \\ 30 & -180 & 180 \end{bmatrix}$$

The spreadsheet for this is shown in the following table.

| Sub mult_main: Example Input and Output | | | | | | | | | |
|---|---|---|---|---|---|---|---|---|---|
| Input | | | | | | | Output | | |
| 3 | 3 | 3 | | | | | | | |
| 1 | 0.5 | 0.333333 | | 9 | -36 | 30 | 1 | 0 | 0 |
| 0.5 | 0.333333 | 0.25 | | -36 | 192 | -180 | 0 | 1 | 0 |
| 0.333333 | 0.25 | 0.2 | | 30 | -180 | 180 | 0 | 0 | 1 |

Two points of interest are as follows.

- Because $[A][B] = \begin{bmatrix} 1 & 0 & 0 \\ 0 & 1 & 0 \\ 0 & 0 & 1 \end{bmatrix}$, $[B]$ is the inverse of $[A]$.

- $[A]$ is called a Hilbert matrix. Each element in its inverse is an integer.

## A Program to Solve Ax = b

This program centers around **Equation Step** to change matrix $[A \mid b]$ to $[I \mid x]$.

$$A(i,j) \ = \ A(i,j) \ - \ A(p,j)\frac{A(i,p)}{A(p,p)}$$

Equation Step

To use **Equation Step**, $A(p,p)$ must not equal zero. The program includes code that reorders the rows to achieve this. The equations and syntax for this code are in the following table.

| | VBA Code to Find an Equation wherein A( p,p ) Is Not Equal to Zero | |
|---|---|---|
| Line 1 | For i = p To N | |
| Line 2 | If Abs(A(i, p)) < tol Then A(i, p) = 0 | Line 2: Abs(A(i, p)) is syntax for $\lvert A(i,p) \rvert$. This puts a tolerance on zero. |
| Line 3 | If Abs(A(i, p)) <> 0 Then Exit For | Line 3: An equation with $A(p,p) \neq 0$ has been found. The For-Next loop is exited and control branches to Line 8. |
| Line 4 | If i = N Then | Lines 4-6: There is no equation having $A(p,p) \neq 0$. This whole subprogram is exited because there is no solution. |
| Line 5 | fail = 1: Exit Sub | |
| Line 6 | End If | |
| Line 7 | Next i | |
| Line 8 | For j = 1 To N + 1 | Lines 8-12: Equations p and i are interchanged even if p = i. This is done to simplify the coding. |
| Line 9 | temp = A(p, j) | |
| Line 10 | A(p, j) = A(i, j) | |
| Line 11 | A(i, j) = temp | |
| Line 12 | Next j | |

The following table discusses the code used to implement **Equation Step**.

When the element in any column $p$ is being zeroed, each element in that row is modified by Equation Step:

$$A(i,j) \;=\; A(i,j) \;-A(p,j)\frac{A(i,p)}{A(p,p)} \qquad \text{Equation Step}$$

For example, in the row where $i = (p + 2)$, each element becomes

$$A(p+2,j)=A(p+2,j)-A(p,j)\frac{A(p+2,p)}{A(p,p)}$$

In that same row, when $j = p$,

$$A(p+2,p)=A(p+2,p)-A(p,p)\frac{A(p+2,p)}{A(p,p)} \;=\; 0$$

This is the desired result.

| The following code implements Step 1 when zeroing the elements below the diagonal. | The following code implements Step 2 when zeroing the elements above the diagonal. |
| --- | --- |
| ``` For p = 1 To N - 1 Step 1   For i = p + 1 To N     M(i) = A(i, p) / A(p, p)     For j = 1 To N + 1       A(i, j) = A(i, j) - M(i) * A(p, j)     Next j   Next i Next p ``` | ``` For p = N To 2 Step -1   For i = 1 To p - 1     M(i) = A(i, p) / A(p, p)     For j = p To N + 1       A(i, j) = A(i, j) - M(i) * A(p, j)     Next j   Next i Next p ``` |

We can now present the program to solve $Ax = b$. The following table contains the listing. Annotations show the following.

- The setup of the matrix $[A \mid b]$.
- Step 1, which includes the logic to avoid $A(p, p) = 0$.
- Step 2, which includes the test that if $A(n, n) = 0$ then there is no solution.
- Step 3, which creates the final identity matrix.

## Program to Solve Linear Equations

| Code page 1 of 2 | Code page 2 of 2 |
|---|---|

```
Sub lineq_main()
 Dim A(50, 51), xd(50), bd(50), ad(50, 51), bdck(50)
 N = Cells(1, 1)                          ' read size A
 For i = 1 To N
   For j = 1 To N + 1
     A(i, j) = Cells(i + 1, j)        ' read A augmented
   Next j
 Next i
 For i = 1 To N
    For j = 1 To N
      ad(i, j) = A(i, j)
    Next j
      bd(i) = A(i, N + 1)
 Next i
 Call lineq(ad, bd, xd, N, fail)
 If fail = 1 Then
   Cells(N + 3, 1) = "no solution"
 Else
   For i = 1 To N
      bdck(i) = 0
      For j = 1 To N
       bdck(i) = bdck(i) + ad(i, j) * xd(j)
      Next j
   Next i
   For i = 1 To N
     Cells(i + N + 3, 1) = xd(i)
     Cells(i + N + 3, 3) = bdck(i)
   Next i
   Cells(N + 3, 1) = "    X": Cells(N + 3, 3) = "     b"
 End If
End Sub ' lineq_main
```

```
Sub lineq(ad, bd, x, N, fail)
 Dim M(50), A(50, 51)
  fail = 0: tol = 0.000001
  If N = 1 Then
    If Abs(ad(1, 1)) < tol Then ad(1, 1) = 0
    If Abs(ad(1, 1)) = 0 Then
      fail = 1
    Else
      x(1) = bd(1) / ad(1, 1)
    End If
    Exit Sub
  End If
  For i = 1 To N                           ' Setup A matrix
    For j = 1 To N
      A(i, j) = ad(i, j)
    Next j
      A(i, N + 1) = bd(i)
  Next i
 For p = 1 To N - 1 Step 1                  ' Step 1 start
                            ' Logic to avoid A(p,p) = 0   start
    For i = p To N
      If Abs(A(i, p)) < tol Then A(i, p) = 0
      If Abs(A(i, p)) <> 0 Then Exit For
      If i = N Then
        fail = 1: Exit Sub
      End If
    Next i
    For j = 1 To N + 1
      temp = A(p, j): A(p, j) = A(i, j): A(i, j) = temp
    Next j
                            ' Logic to avoid A(p,p) = 0   end
    For i = p + 1 To N
       M(i) = A(i, p) / A(p, p)
       For j = 1 To N + 1
        A(i, j) = A(i, j) - M(i) * A(p, j)
       Next j
    Next i
  Next p                                    ' Step 1 end
  For p = N To 2 Step -1                     ' Step 2 start
    For i = 1 To p - 1
     If Abs(A(p, p)) < tol Then A(p, p) = 0
     If Abs(A(p, p)) = 0 Then
       fail = 1: Exit Sub               ' Test for A(n,n) = 0
     End If
      M(i) = A(i, p) / A(p, p)
      For j = p To N + 1
       A(i, j) = A(i, j) - M(i) * A(p, j)
      Next j
    Next i
  Next p                                    ' Step 2 end
  For i = 1 To N
     x(i) = A(i, N + 1) / A(i, i)            ' Step 3
  Next i
 End Sub  ' lineq
```

65

Sub **lineq_main** is used to solve the problem:

$$\begin{bmatrix} 1 & 1/2 & 1/3 \\ 1/2 & 1/3 & 1/4 \\ 1/3 & 1/4 & 1/5 \end{bmatrix} \begin{bmatrix} x_1 \\ x_2 \\ x_3 \end{bmatrix} = \begin{bmatrix} 1 \\ 1 \\ 1 \end{bmatrix}$$

| Spreadsheet Input and Output for Sub lineq_main | | | | | | | |
|---|---|---|---|---|---|---|---|
| Input | | | | | Output | | |
| 3 | | | | | X | | b |
| 1 | 0.5 | 0.333333 | 1 | | | 3 | 1 |
| 0.5 | 0.333333 | 0.25 | 1 | | | -24 | 1 |
| 0.333333 | 0.25 | 0.2 | 1 | | | 30 | 1 |

The output includes a column labeled **b**. After x has been computed, the program computes Ax to show the accuracy of the solution.

## A Program to Compute the Inverse of a Matrix

This program operates like the program to solve $Ax = b$.

- Form a matrix. $\left[ A \mid I \right]$.
- Use row operations to change this matrix to $\left[ I \mid A^{-1} \right]$.

Since the process is the same except for matrix size, no new syntax is needed. The following table contains the listing with the following annotations.

- The setup of the matrix $\left[ A \mid I \right]$.
- Step 1, which includes the logic to avoid $A(p, p) = 0$.
- Step 2, which includes the test that if $A(n, n) = 0$, then there is no solution.
- Step 3, which creates the final identity matrix.

## Program to Invert a Matrix

| Code page 1 of 2 | Code page 2 of 2 |
|---|---|

```
Sub inverse_main()
  Dim A(20, 50), Ain(20, 20), ainv(20, 20), aainv(20, 20)
  N = Cells(1, 1)                    ' read size
  For i = 1 To N
    For j = 1 To 2 * N
      A(i, j) = 0
    Next j
  Next i
  For i = 1 To N
    For j = 1 To N
      Ain(i, j) = Cells(i + 1, j)         ' read Ain
      A(i, j) = Ain(i, j)
    Next j
  Next i
  For i = 1 To N
    A(i, i + N) = 1                  ' The identity matrix
  Next i
  Call inverse(A, ainv, N, fail)
  If fail = 1 Then
    Cells(N + 3, 1) = "no solution": End
  End If
  For i = 1 To N
   For j = 1 To N
    Cells(N + 3 + i, j) = Application.Round(ainv(i, j), 3)
   Next j
  Next i
  Call mult(Ain, ainv, aainv, N, N, N)
    For i = 1 To N
     For j = 1 To N
       Cells(2 * N + 5 + i, j) = aainv(i, j)
     Next j
    Next i
  Cells(N + 3, 1) = "Ainv": Cells(2 * N + 5, 1) = "A*Ainv"
End Sub  ' inverse_main
Sub mult(A, B, AB, M, L, N)
  For i = 1 To M
    For j = 1 To N
      AB(i, j) = 0
      For k = 1 To L
        AB(i, j) = AB(i, j) + A(i, k) * B(k, j)
      Next k
    Next j
  Next i
End Sub  ' mult
```

```
Sub inverse(A, ainv, N, fail)
Dim M(50)
  tol = 0.000001: fail = 0
  For p = 1 To N - 1 Step 1              ' Step 1 start
                    ' Logic to avoid A(p,p)= 0 start
    For i = p To N
      If Abs(A(i, p)) < tol Then A(i, p) = 0
      If Abs(A(i, p)) <> 0 Then Exit For
      If i = N Then
        fail = 1: Exit Sub
      End If
    Next i
    For j = 1 To N + N
      temp = A(p, j): A(p, j) = A(i, j): A(i, j) = temp
    Next j
                    ' Logic to avoid A(p,p)= 0 end
    For i = p + 1 To N
      M(i) = A(i, p) / A(p, p)
      For j = 1 To N + N
       A(i, j) = A(i, j) - M(i) * A(p, j)
      Next j
    Next i
  Next p                        ' Step 1 end
  For p = N To 2 Step -1             ' Step 2 start
    For i = 1 To p - 1
      If Abs(A(p, p)) < tol Then A(p, p) = 0
      If Abs(A(p, p)) = 0 Then  ' Test for A(n,n) = 0
        fail = 1: Exit Sub
      End If
       M(i) = A(i, p) / A(p, p)
      For j = p To N + N
       A(i, j) = A(i, j) - M(i) * A(p, j)
      Next j
    Next i
  Next p                        ' Step 2 end
  For i = 1 To N                 ' Step 3 Start
    For j = N + 1 To N + N
      A(i, j) = A(i, j) / A(i, i)
    Next j
    A(i, i) = 1
  Next I                         ' Step3 End
  For i = 1 To N
    For j = N + 1 To N + N
      ainv(i, j - N) = A(i, j)
    Next j
  Next i
End Sub ' inverse
```

Sub inverse_main is used to solve the following problem.

We are given $A = \begin{bmatrix} 1 & 1/2 & 1/3 \\ 1/2 & 1/3 & 1/4 \\ 1/3 & 1/4 & 1/5 \end{bmatrix}$. Find $A^{-1}$.

| Spreadsheet Input and Output for Sub inverse_main | | | | | | | | |
|---|---|---|---|---|---|---|---|---|
| Input | | | Output | | | | | |
| 3 | | | Ainv | | | A*Ainv | | |
| 1 | 0.5 | 0.333333 | 9 | -36 | 30 | 1 | 7.11E-15 | 7.11E-15 |
| 0.5 | 0.333333 | 0.25 | -36 | 192 | -180 | 8.88E-16 | 1 | 0 |
| 0.333333 | 0.25 | 0.2 | 30 | -180 | 180 | 1.78E-15 | 0 | 1 |

The output includes a matrix labeled A*Ainv. After $A^{-1}$ has been computed, the program computes $A * A^{-1}$ to show the accuracy of the solution.

# Chapter 4: Polynomials

This chapter is about polynomials. The following is an example.

$$F(x) = x^3 + 8x^2 + 32x + 300$$

In general,

$$F(x) = c_n x^n + c_{n-1} x^{n-1} + \cdots + c_1 x + c_0$$

- $n$ is a positive integer.
- x is the variable which may be complex.
- The coefficients ( $c_n$, etc. ) are constants and are real numbers.

This chapter begins with a discussion of where some polynomials come from. The following table shows application of Newton's law to a spring-mass system. It also shows a time response when the spring is released from a compressed position.

| 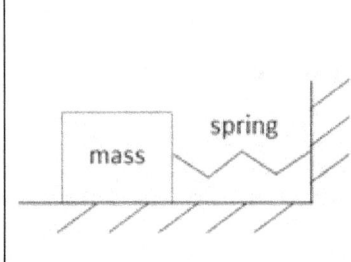 | • The spring force equals $k \times Z$, where $Z$ is the displacement of the mass, and $k$ is a known constant.<br>• The friction force equals $d \times V$, where $V$ is the velocity of the mass, and $d$ is a known constant. Note that $V = \dfrac{\Delta Z}{\Delta t}$, i.e., change in $Z$ due to change in time $\Delta t$.<br>• The mass force equals $m \times \dfrac{\Delta V}{\Delta t}$, where $\dfrac{\Delta V}{\Delta t}$ is change in $V$ due to $\Delta t$.<br>• Newton's law states that the sum of the forces equals zero.<br>$$m \times \frac{\Delta V}{\Delta t} + d \times \frac{\Delta Z}{\Delta t} + k \times Z = 0$$ |
|---|---|
| For $m = 0.5$, $d = 2$ and $k = 80$, the adjacent figure shows the time response when the spring is released from a compressed position. | 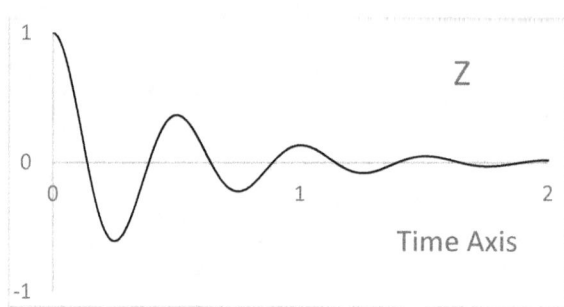 |

From calculus, the solution to the spring-mass system has the form $Z = Ke^{\alpha t}$.

- α is a constant and a function of $m$, $d$ and $k$. (α may be complex).
- $t$ is time.
- K is the initial value of $Z$.

The following discussion shows that α is a zero of a polynomial. In chapter 2 we showed that when $Z = Ke^{\alpha t}$, then $\dfrac{\Delta Z}{\Delta t} = \alpha Z$. Since $V = \dfrac{\Delta Z}{\Delta t}$, then $V = \alpha Z$. Further, $\dfrac{\Delta V}{\Delta t} = \dfrac{\Delta \alpha Z}{\Delta t} = \alpha V$.

Therefore, $\dfrac{\Delta V}{\Delta t} = \alpha^2 Z$. Let's repeat Newton's Law for the spring-mass system.

$$m\frac{\Delta V}{\Delta t} + d\frac{\Delta Z}{\Delta t} + k\,Z = 0$$

If we substitute $\dfrac{\Delta V}{\Delta t} = \alpha^2 Z$ and $\dfrac{\Delta Z}{\Delta t} = \alpha Z$, we get $m\alpha^2 Z + d\alpha Z + k\,Z = 0$.

In factored form, $$\left(m\alpha^2 + d\alpha + k\right)Z = 0$$

In chapter 3, we showed that for non-zero $Z$, $\left(m\alpha^2 + d\alpha + k\right) = 0$. This shows the reason and the importance of the equation that sets a polynomial equal to zero. In fact, the equation $F(x) = 0$ is given the special name, the *Characteristic* equation.

$$F(x) = c_n x^n + c_{n-1} x^{n-1} + \bullet\bullet\bullet + c_1 x + c_0 = 0.$$

The values of $x$ that make $F(x) = 0$ are called the *zeros* of $F(x)$.

Let's plug in some numbers. If $m = 0.5$, $d = 2$, and $k = 80$, then $F(\alpha) = \left(0.5\alpha^2 + 2\alpha + 80\right) = 0$. The Quadratic formula yields the two solutions.

$$\alpha_1 = \frac{-2 + \sqrt{2^2 - 4*0.5*80}}{2*0.5} = -2 + j12.5 \text{ and } \alpha_2 = -2 - j12.5.$$

These are a complex conjugate pair. They correspond to the time response shown on the previous page. The following is a plot of $F(\alpha)$ for real values of α.

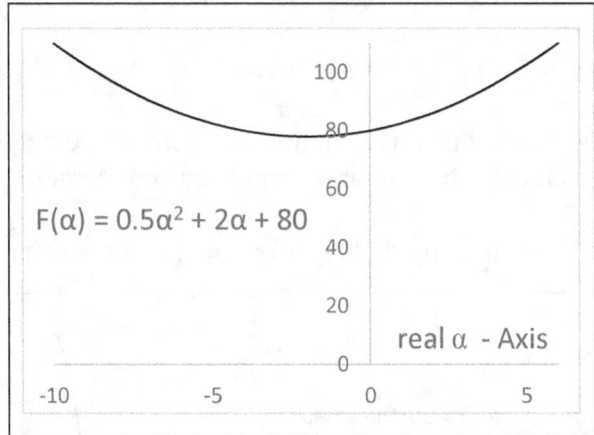

This plot does not cross the real α-axis. In chapter 1, we showed that at complex α's, $F(\alpha)$ goes to zero in the complex plane rather than on the real axis.

**Case with Increased Friction:** The following table shows the response.

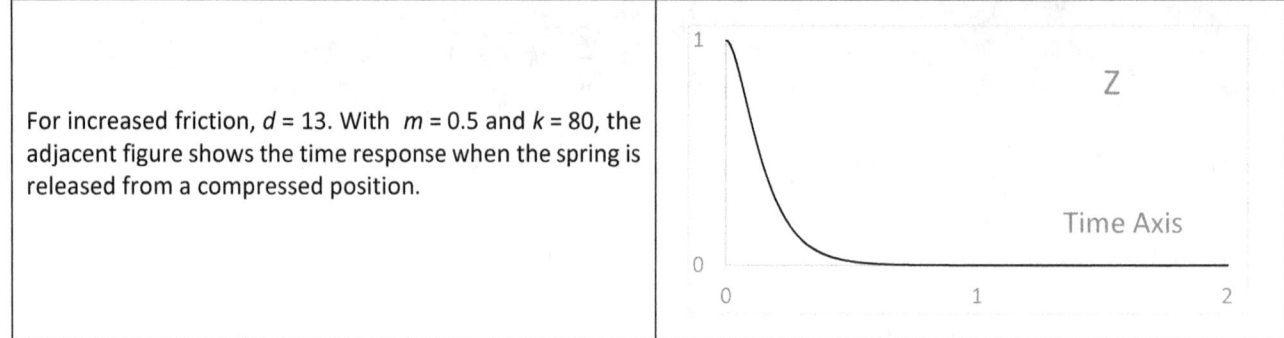

For increased friction, $d = 13$. With $m = 0.5$ and $k = 80$, the adjacent figure shows the time response when the spring is released from a compressed position.

For this case, the polynomial equation is $F(\alpha) = (0.5\alpha^2 + 13\alpha + 80) = 0$. The Quadratic formula yields the two solutions.

$$\alpha_1 = \frac{-13 + \sqrt{13^2 - 4*0.5*80}}{2*0.5} = -10 \text{ and } \alpha_2 = -16.$$

These are two real-valued zeros. The following is a plot of this $F(\alpha)$ for real values of α. Here, because the zeros are real, the α-axis is crossed twice.

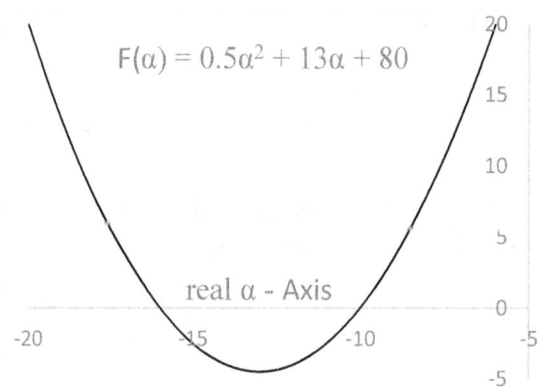

$F(\alpha) = 0.5\alpha^2 + 13\alpha + 80$

When the spring-mass system has positive friction, the zeros have negative real parts. They are in the left-half plane (LHP). This means that the time response is convergent.

**Case with No Friction:** The following table shows how the mass responds for this case.

For no friction, $d = 0$. With $m = 0.5$ and $k = 80$, the adjacent figure shows the time response when the spring is released from a compressed position. Note: It is not realistic for $d$ to be exactly zero, but it can be very small.

For this case, $F(\alpha) = (0.5\alpha^2 + 80) = 0$. The zeros are $\alpha_1 = j12.65$ and $\alpha_2 = -j12.65$. They are not in the left-half plane (LHP) or in the right-half plane (RHP). Here, the time response is oscillatory.

**Case with Negative Friction:** If it were possible to have negative friction, the response would be as shown in the following table.

| | |
|---|---|
| For negative friction, $d = -2$. With $m = 0.5$ and $k = 80$, the adjacent figure shows the time response when the spring is released from a compressed position. Note: Though negative friction is not realistic, many complicated real systems can experience divergence like this. | 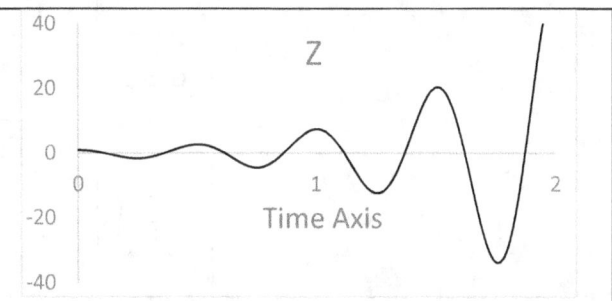 |

For this case, $F(\alpha) = (0.5\alpha^2 - 2\alpha + 80) = 0$. The zeros are $\alpha_1 = +2 + j12.5$ and $\alpha_2 = +2 - j12.5$. These are in the RHP. The time response is divergent. Hence it is important to classify zeros according to LHP or RHP.

This chapter started with the third-order polynomial $F(x) = x^3 + 8x^2 + 32x + 300$. The zeros of this polynomial are those values of x that satisfy the equation: $F(x) = x^3 + 8x^2 + 32x + 300 = 0$. These zeros may be real and/or complex, and of course, RHP and/or LHP. Finding zeros of polynomials other than second-order is not easy. But an inspection of the polynomial itself can show a lot about its zeros.

The above polynomial has three zeros. If x is a large and positive real number, F(x) will be a large and positive real number. Also, if x is a large and negative real number, F(x) will be a large and negative real number. Hence, there is at least one real value of x for which F(x) = 0. In fact, all third-order polynomials have at least one real-valued zero.

This can be seen another way. F(x) can be factored in only two ways;

    1) $F(x) = (x - \omega_1)(x - \omega_2)(x - \omega_3)$ where $\omega_1$, $\omega_2$ and $\omega_3$ are real numbers;
    2) $F(x) = (x - \omega_1)(x - \alpha + j\beta)(x - \alpha - j\beta)$ where $\omega_1$, $\alpha$ and $\beta$ are real numbers.

Either way, there is always one real zero. There may be three real zeros, but not two. It's important to know if a zero is in the RHP or LHP. The following table shows the possible groupings for the case with all real zeros.

| Possible Locations for $\omega_1$, $\omega_2$ and $\omega_3$ | | |
|---|---|---|
| Group | RHP | LHP |
| 1 | 3 | 0 |
| 2 | 2 | 1 |
| 3 | 1 | 2 |
| 4 | 0 | 3 |

The following table shows the groupings for the other case.

| | Possible Locations for $\omega_1$ and $\alpha \pm j\beta$ | | | |
|---|---|---|---|---|
| | $\omega_1$ | | $\alpha \pm j\beta$ | |
| Group | RHP | LHP | RHP | LHP |
| 1 | 1 | 0 | 1 | 0 |
| 2 | 1 | 0 | 0 | 1 |
| 3 | 0 | 1 | 1 | 0 |
| 4 | 0 | 1 | 0 | 1 |

For all third-order polynomials, these are the eight possible groupings.

Repeating our third-order polynomial $F(x) = x^3 + 8x^2 + 32x + 300$, all the coefficients are positive. Therefore, for all real $x \geq 0$, the minimum value is $F(x) = 300$. That means that the one real zero it must have, must be in the LHP.

For negative real $x$, $F(-x) = -x^3 + 8x^2 - 32x + 300$. There are three sign changes. Since the terms are not all additive, the value of $F(-x)$ can increase and decrease as $x$ varies. But are these oscillations large enough so that $F(x)$ has more than one negative real zero? By looking only at the coefficients, we can't tell the difference between the two cases that are shown in the following table.

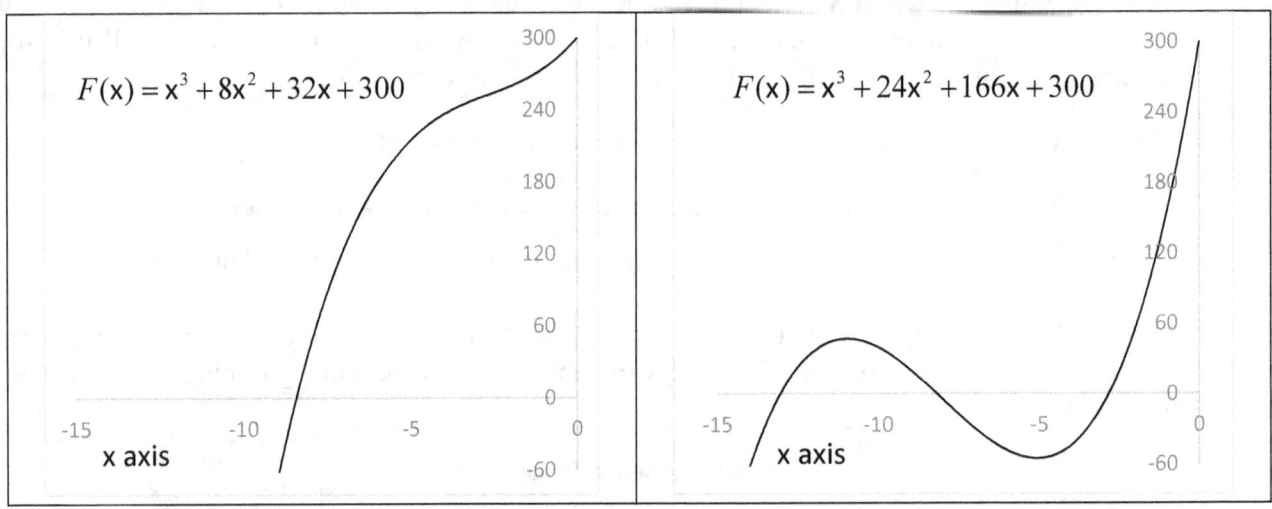

Coefficients of a polynomial can be examined in another way to determine the location of its zeros. This way uses the Routh array, which is an array of numbers that are derived from the polynomial coefficients. The derivation of this array is in Schwarz and Friedland[1].

---

[1] Schwarz, R., and B. Friedland, *Linear Systems* (New York: McGraw-Hill Book Company, 1965).

For the polynomial $F(x) = x^3 + k_2 x^2 + k_3 x + k_4$, the Routh array is given in the following table.

**Routh Array For**
$$F(x) = x^3 + k_2 x^2 + k_3 x + k_4$$

$$\begin{bmatrix} 1 & k_3 \\ k_2 & k_4 \\ k_3 - \dfrac{k_4}{k_2} & \\ k_4 & \end{bmatrix}$$

Routh proved that the number of zeros in the RHP equals the number of sign changes in the first column of the array. The following table shows the array for $F(x) = x^3 + 8x^2 + 32x + 300$.

**Routh Array For**
$$F(x) = x^3 + 8x^2 + 32x + 300$$

$$\begin{bmatrix} 1 & 32 \\ 8 & 300 \\ -5.5 & \\ 300 & \end{bmatrix}$$

Since there are two sign changes in the first column, there are two zeros in the RHP. Since these can't be real zeros, they must be complex. We now conclude that the polynomial has one real zero in the LHP, and a complex conjugate pair in the RHP. Eight possibilities have been reduced to this single result. For completeness, the polynomial factors as follows:

$$F(x) = \underbrace{(x + 8.43)}_{\text{LHP Real}} \underbrace{(x - 0.214 + j6)(x - 0.214 - j6)}_{\text{RHP Complex}}$$

This analysis involved two steps:

Step 1: We determined the number of sign changes in $F(x)$ and $F(-x)$. This step was the application of Descartes' *Rule of Signs*. This is defined in the following table.

---

**Descartes' Rule of Signs for a Polynomial F(x)**

---

This is for a polynomial with real coefficients written in ascending or descending order.

1) The number of real zeros in the right-half plane is either:
> • equal to the number of variations in the signs of the coefficients of F(x);
> or, • less than that by an even whole number.

2) The number of real zeros in the left-half plane is either:
> • equal to the number of variations in the signs of the coefficients of F(-x);
> or, • less than that by an even whole number.

Notes:
• Zero coefficients are not counted.
• Descartes' rule says nothing directly about complex zeros. But inferences about them can be made.

---

Examples: • $F(x) = x^3 + 8x^2 + 32x - 300$. The number of sign changes is one.

• $F(x) = x^4 - 5x^2 + 10$. The number of sign changes is two.

---

For our third-order polynomial, we concluded that $F(x)$ has no real zeros in the RHP, and has either three or one real zero in the LHP. The following table summarizes these results.

| RHP test:<br>$F(x) = x^3 + 8x^2 + 32x + 300$<br>No sign change means no real zeros in RHP.<br><br>LHP test:<br>$F(-x) = -x^3 + 8x^2 - 32x + 300$<br>Three sign changes means either three or one real zero in LHP. | | | |
|---|---|---|---|

| Descartes' *Rule of Signs* Yields These Possibilities | | | |
|---|---|---|---|
| | Number of Real Zeros | | |
| Group | RHP | LHP | Number of Complex |
| 1 | 0 | 3 | 0 |
| 2 | 0 | 1 | 2 |

At this point, we only know how many complex zeros are in each group.

Step 2: We formed the Routh array. That array eliminated group 1. We were finally able to tell that F(x) has one real zero in the LHP, and one complex pair in the RHP.

The following table shows how to construct a Routh array for

$$F(x) = x^n + k_2 x^{n-1} + k_3 x^{n-2} + \cdots + k_n x + k_{n+1}$$

| | | | | | |
|---|---|---|---|---|---|
| • Entries in the first two rows come directly from the polynomial.<br>• Entries below row 2 are computed from entries in the previous two rows.<br>• The array has (N+1) rows, but it is not a square. | | | | | |
| | $j = 1$ | $j = 2$ | $j = 3$ | $j = 4$ | ••• |
| $i = 1$ | $a_{11} = 1$ | $a_{12} = k_3$ | $a_{13} = k_5$ | $a_{14} = k_7$ | ••• |
| $i = 2$ | $a_{21} = k_2$ | $a_{22} = k_4$ | $a_{23} = k_6$ | $a_{24} = k_8$ | ••• |
| •<br>•<br>• | $i$ goes from 3 to $(N+1)$. For each $i$, $j = 1$ to $X^\Delta$ : $$a_{i,j} = a_{i-2,j+1} - a_{i-1,j+1}\frac{a_{i-2,1}}{a_{i-1,1}}$$ $^\Delta X$ varies with row. A row ends when the rest of the row becomes all zeros. | | | | |
| Important note: If $a_{i-1,1} = 0$, it is valid to replace it with a very small number ( e.g., 0.0001 ), and then continue forming the array. This will be demonstrated in examples that follow. | | | | | |

## Examples Using Descartes' Rule of Signs and the Routh Array

• Polynomial 1:   $F(x) = x^4 + x^3 + 2x^2 + 2x + 5$. The zeros of a polynomial can be grouped as real or complex, and in the RHP or LHP. For a fourth-order polynomial, there are fourteen possible groups.

Step 1: Descartes' Rule of Signs.

| | | Descartes' *Rule of Signs* Yields These Possibilities | | |
|---|---|---|---|---|
| RHP test:<br>$\qquad F(x) = x^4 + x^3 + 2x^2 + 2x + 5$<br>No sign change means no real zeros in RHP. | | | Number of Real Zeros | |
| LHP test:<br>$\qquad F(-x) = x^4 - x^3 + 2x^2 - 2x + 5$<br>Four sign changes means there are at most, four real zeros. Because of the possibility of complex zeros, there may be four, two or no real zeros in LHP. | | Group | RHP | LHP | Number of Complex |
| | | 1 | 0 | 4 | 0 |
| | | 2 | 0 | 2 | 2 |
| | | 3 | 0 | 0 | 4 |

Step 2: Given the following Routh array for a fourth-order polynomial, determine how many zeros are in each plane (RHP and LHP).

$$\begin{bmatrix} 1 & k_3 & k_5 \\ k_2 & k_4 & \\ d_1 = k_3 - k_4\dfrac{1}{k_2} & d_2 = k_5 & \\ k_4 - d_2\dfrac{k_2}{d_1} & & \\ k_5 & & \end{bmatrix}$$

| The Routh Array | | | | Zeros from Column 1 of the Routh Array | |
|---|---|---|---|---|---|
| column 1 | column 2 | column 3 | | In the Right-Half Plane | In the Left-Half Plane |
| 1 | 2 | 5 | | | |
| 1 | 2 | | | | |
| 0.0001 | 5 | | | 2 | 2 |
| -49998 | | | | | |
| 5 | | | | Group 1 is ruled out by the Routh array. | |

Summary of the Possibilities

| Group | Number of Real Zeros | | Number of Complex | |
|---|---|---|---|---|
| | RHP | LHP | RHP | LHP |
| 2 | 0 | 2 | 2 | 0 |
| 3 | 0 | 0 | 2 | 2 |

For this polynomial, there are two possible groups out of the original fourteen. The actual zeros are in group 3.

$$F(x) = \underbrace{(x-.58+j1.35)(x-.58-j1.35)}_{\text{RHP Complex}}\underbrace{(x+1.08+j1.07)(x+1.08-j1.07)}_{\text{LHP Complex}}$$

• Polynomial 2: $F(x) = x^5 + 2.5x^3 + 1.5$. This polynomial has several zero coefficients. For a fifth-order polynomial, the zeros can be grouped in twenty possible ways.

Step 1: Descartes' Rule of Signs.

| RHP test:<br>$$F(x) = x^5 + 2.5x^3 + 1.5$$<br>No sign changes mean no real zeros in RHP.<br><br>LHP test:<br>$$F(-x) = -x^5 - 2.5x^3 + 1.5$$<br>One sign change means exactly one real zero in LHP. | | Descartes' *Rule of Signs* Yields These Possibilities | | | |
| --- | --- | --- | --- | --- | --- |
| | | | Number of Real Zeros | | |
| | | Group | RHP | LHP | Number of Complex |
| | | 1 | 0 | 1 | 4 |

Step 2: Given the following Routh array for a fifth-order polynomial, determine how many zeros are in each plane (RHP and LHP).

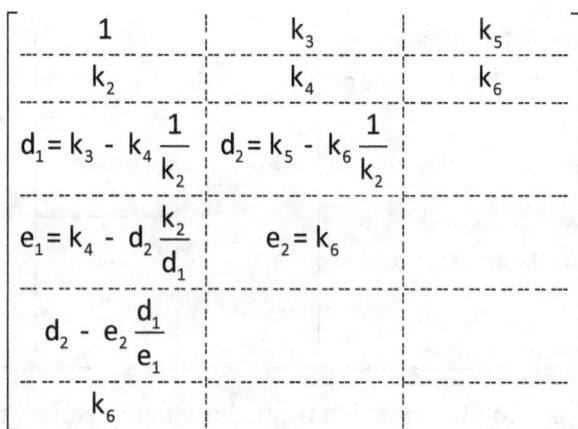

| The Routh Array | | | | Zeros from Column 1 of the Routh Array | |
| --- | --- | --- | --- | --- | --- |
| column 1 | column 2 | column 3 | | In the Right-Half Plane | In the Left-Half Plane |
| 1 | 2.5 | 0 | | | |
| 0.0001 | 0 | 1.5 | | 2 | 3 |
| 2.5 | -15000 | | | | |
| 0.6 | 1.5 | | | | |
| -15006.3 | | | | | |
| 1.5 | | | | | |

## Summary of the Possibilities

| | Group | Number of Real Zeros | | Number of Complex | |
|---|---|---|---|---|---|
| | | RHP | LHP | RHP | LHP |
| | 1 | 0 | 1 | 2 | 2 |

The zeros of this polynomial are in only one group out of the original twenty. The actual zeros are

$$F(x) = \underbrace{(x+0.78)}_{\text{LHP Real}}\underbrace{(x-0.5+j0.7)(x-0.5-j0.7)}_{\text{RHP Complex}}\underbrace{(x+0.11+j1.6)(x+0.11-j1.6)}_{\text{LHP Complex}}$$

---

- Polynomial 3: $F(x) = x^6 - 1.5x^5 + 1.5x^3 - 2.5x^2 + 3x - 4.5$. The zeros can be grouped in thirty ways.

Step 1: Descartes' Rule of Signs.

<table>
<tr><td>

RHP test:

$F(x) = x^6 - 1.5x^5 + 1.5x^3 - 2.5x^2 + 3x - 4.5$

Five sign changes means there can be five, three or one real zero in the RHP.

LHP test:

$F(-x) = x^6 + 1.5x^5 - 1.5x^3 - 2.5x^2 - 3x - 4.5$

One sign change means exactly one real zero in LHP.

</td><td>

Descartes' *Rule of Signs* Yields These Possibilities

| | Number of Real Zeros | | |
|---|---|---|---|
| Group | RHP | LHP | Number of Complex |
| 1 | 5 | 1 | 0 |
| 2 | 3 | 1 | 2 |
| 3 | 1 | 1 | 4 |

</td></tr>
</table>

Step 2: Given the following Routh array, determine how many zeros are in each plane (RHP and LHP).

$$\begin{bmatrix} 1 & k_3 & k_5 & k_7 \\ k_2 & k_4 & k_6 & \\ d_1 = k_3 - k_4\dfrac{1}{k_2} & d_2 = k_5 - k_6\dfrac{1}{k_2} & d_3 = k_7 & \\ e_1 = k_4 - d_2\dfrac{k_2}{d_1} & e_2 = k_6 - d_3\dfrac{k_2}{d_1} & & \\ f_1 = d_2 - e_2\dfrac{d_1}{e_1} & f_2 = k_7 & & \\ e_2 - f_2\dfrac{e_1}{f_1} & & & \\ k_7 & & & \end{bmatrix}$$

| The Routh Array | | | | | Zeros from Column 1 of the Routh Array | |
|---|---|---|---|---|---|---|
| column 1 | column 2 | column 3 | column 4 | | In the Right-Half Plane | In the Left-Half Plane |
| 1 | 0 | -2.5 | -4.5 | | | |
| -1.5 | 1.5 | 3 | | | 3 | 3 |
| 1 | -0.5 | -4.5 | | | | |
| 0.75 | -3.75 | | | | | |
| 4.5 | -4.5 | | | | Group 1 is ruled out by the Routh array. | |
| -3 | | | | | | |
| -4.5 | | | | | | |

## Summary of the Possibilities

| Group | Number of Real Zeros | | Number of Complex | |
|---|---|---|---|---|
| | RHP | LHP | RHP | LHP |
| 2 | 3 | 1 | 0 | 2 |
| 3 | 1 | 1 | 2 | 2 |

The zeros can be in two of the original thirty groups. The actual zeros are in group 3.

$$F(x) = \underbrace{(x-1.54)}_{\text{RHP Real}}\underbrace{(x+1.45)}_{\text{LHP Real}}\underbrace{(x-0.92+j0.93)(x-0.92-j0.93)}_{\text{RHP Complex}}\underbrace{(x+0.22+j1.07)(x+0.22-j1.07)}_{\text{LHP Complex}}$$

For the next example, we have to discuss a rare occurrence when using the Routh array. It is theoretically possible for a polynomial to have zeros that are mirror images of each other. For example $F(x) = (x+\alpha)(x-\alpha)\cdots$, or $F(x) = (x+j\beta)(x-j\beta)\cdots$. Because of tolerances, these cases do not appear in reality. But they must be discussed because they do appear in textbooks. As we will see, a mirror pair results in a row of zeros in the Routh array. What sign should be put on a zero in column 1 of the array? Wolovich[2] shows a complicated way to handle this. A simple way is just to ignore the row of zeros when counting the sign changes. This simple way works for a single pair, but not for more than one pair. This limitation is not serious because of the rarity of occurrence.

[2] Wolovich,W., *Automatic Control Systems* (Orlando, Florida: Saunders College Publishing, 1994).

• Polynomial 4: $F(x) = x^4 - x^3 - 5x^2 - x - 6$. In this example, the polynomial has a *mirror pair* of zeros on the $j$-axis. Of course, we don't know this at the beginning of the analysis.

Step 1: Descartes' Rule of Signs.

<table>
<tr><td rowspan="6">RHP test:<br><br>$$F(x) = x^4 - x^3 - 5x^2 - x - 6$$<br>One sign change means one real zero in RHP.<br><br>LHP test:<br>$$F(-x) = x^4 + x^3 - 5x^2 + x - 6$$<br>Three sign changes mean either three or one real zero in LHP.</td><td colspan="4">Descartes' *Rule of Signs* Yields These Possibilities</td></tr>
<tr><td></td><td colspan="2">Number of Real Zeros</td><td></td></tr>
<tr><td>Group</td><td>RHP</td><td>LHP</td><td>Number of Complex</td></tr>
<tr><td>1</td><td>1</td><td>3</td><td>0</td></tr>
<tr><td>2</td><td>1</td><td>1</td><td>2</td></tr>
</table>

Step 2: Using the Routh array for a fourth-order polynomial, determine how many zeros are in each plane (RHP and LHP).

<table>
<tr><td colspan="3">The Routh Array</td><td></td><td colspan="2">Zeros from Column 1 of the Routh Array</td></tr>
<tr><td>column 1</td><td>column 2</td><td>column 3</td><td></td><td>In the Right-Half Plane</td><td>In the Left-Half Plane</td></tr>
<tr><td>1</td><td>-5</td><td>-6</td><td></td><td></td><td></td></tr>
<tr><td>-1</td><td>-1</td><td></td><td></td><td>1</td><td>3</td></tr>
<tr><td>-6</td><td>-6</td><td></td><td></td><td></td><td></td></tr>
<tr><td>$0.0001^{\Delta}$</td><td></td><td></td><td></td><td></td><td></td></tr>
<tr><td>-6</td><td></td><td></td><td></td><td></td><td></td></tr>
<tr><td>$^{\Delta}$Ignore this row.</td><td></td><td></td><td></td><td></td><td></td></tr>
</table>

Summary of the Possibilities

<table>
<tr><td rowspan="2">Group</td><td colspan="2">Number of Real Zeros</td><td colspan="2">Number of Complex</td></tr>
<tr><td>RHP</td><td>LHP</td><td>RHP</td><td>LHP</td></tr>
<tr><td>1</td><td>1</td><td>3</td><td>0</td><td>0</td></tr>
<tr><td>2</td><td>1</td><td>1</td><td>0</td><td>2</td></tr>
</table>

The zeros can be in two groups out of the original fourteen. The actual zeros are in group 2.

$$F(x) = \underbrace{(x-3)}_{\text{RHP Real}}\underbrace{(x+2)}_{\text{LHP Real}}\underbrace{(x+j)(x-j)}_{\text{Mirror Complex}}$$

• Polynomial 5: $F(x) = x^5 + 2.5x^4 - 4x^3 - 7x^2 + 3x + 4.5$. This polynomial has a *mirror pair* of real axis zeros. Again, we don't know this at the beginning of the analysis.

Step 1: Descartes' Rule of Signs.

<table>
<tr><td rowspan="8">RHP test:<br><br>$F(x) = x^5 + 2.5x^4 - 4x^3 - 7x^2 + 3x + 4.5$<br><br>Two sign changes mean that there are either two or no real zeros in RHP.<br><br>LHP test:<br><br>$F(-x) = -x^5 + 2.5x^4 + 4x^3 - 7x^2 - 3x + 4.5$<br><br>Three sign changes mean that there are either three or one real zero in LHP.</td><td></td><td colspan="4">Descartes' Rule of Signs Yields These Possibilities</td></tr>
<tr><td></td><td colspan="3">Number of Real Zeros</td></tr>
<tr><td></td><td>Group</td><td>RHP</td><td>LHP</td><td>Number of Complex</td></tr>
<tr><td></td><td>1</td><td>2</td><td>3</td><td>0</td></tr>
<tr><td></td><td>2</td><td>2</td><td>1</td><td>2</td></tr>
<tr><td></td><td>3</td><td>0</td><td>3</td><td>2</td></tr>
<tr><td></td><td>4</td><td>0</td><td>1</td><td>4</td></tr>
</table>

Step 2: From the Routh array, determine how many zeros are in each plane (RHP and LHP).

<table>
<tr><td colspan="3">The Routh Array</td><td></td><td colspan="2">Zeros from Column 1 of the Routh Array</td></tr>
<tr><td>column 1</td><td>column 2</td><td>column 3</td><td></td><td>In the Right-Half Plane</td><td>In the Left-Half Plane</td></tr>
<tr><td>1</td><td>-4</td><td>3</td><td></td><td rowspan="5">2</td><td rowspan="5">3</td></tr>
<tr><td>2.5</td><td>-7</td><td>4.5</td></tr>
<tr><td>-1.2</td><td>1.2</td><td></td></tr>
<tr><td>-4.5</td><td>4.5</td><td></td></tr>
<tr><td>$-0.0001^\Delta$</td><td></td><td></td></tr>
<tr><td>4.5</td><td></td><td></td><td></td><td></td><td></td></tr>
<tr><td>$^\Delta$Ignore this row.</td><td></td><td></td><td></td><td></td><td></td></tr>
</table>

Summary of the Possibilities

<table>
<tr><td rowspan="2">Group</td><td colspan="2">Number of Real Zeros</td><td colspan="2">Number of Complex</td></tr>
<tr><td>RHP</td><td>LHP</td><td>RHP</td><td>LHP</td></tr>
<tr><td>1</td><td>2</td><td>3</td><td>0</td><td>0</td></tr>
<tr><td>2</td><td>2</td><td>1</td><td>0</td><td>2</td></tr>
<tr><td>3</td><td>0</td><td>3</td><td>2</td><td>0</td></tr>
<tr><td>4</td><td>0</td><td>1</td><td>2</td><td>2</td></tr>
</table>

The zeros can be in four of the original twenty groups. The actual zeros are in group 1.

$$F(x) = \underbrace{(x-1.5)}_{\text{RHP Real}}\underbrace{(x-1)(x+1)}_{\text{Mirror Real}}\underbrace{(x+3)}_{\text{LHP Real}}\underbrace{(x+1)}_{\text{LHP Real}}$$

## The Relationships between the Coefficients of a Polynomial and Its Zeros

To demonstrate the pattern of these relationships, we will use a fourth-order polynomial. This polynomial can be factored in three ways:

- $F(x) = (x-\omega_1)(x-\omega_2)(x-\omega_3)(x-\omega_4)$
- $F(x) = (x-\omega_1)(x-\omega_2)(x-\alpha+j\beta)(x-\alpha-j\beta)$
- $F(x) = (x-\alpha_1+j\beta_1)(x-\alpha_1-j\beta_1)(x-\alpha_2+j\beta_2)(x-\alpha_2-j\beta_2)$

| $F(x) = (x-\omega_1)(x-\omega_2)(x-\omega_3)(x-\omega_4)$ | This column describes the relationships. |
|---|---|
| $F(x) = x^4$ $-x^3(\omega_1+\omega_2+\omega_3+\omega_4)$ $+x^2(\omega_1\omega_2+\omega_1\omega_3+\omega_1\omega_4+\omega_2\omega_3+\omega_2\omega_4+\omega_3\omega_4)$ $-x^1(\omega_1\omega_2\omega_3+\omega_1\omega_2\omega_4+\omega_1\omega_3\omega_4+\omega_2\omega_3\omega_4)$ $+x^0(\omega_1\omega_2\omega_3\omega_4)$ | $F(x) = x^4$ $-x^3(\text{sum of the zeros})$ $+x^2(\text{sum of products, each from two zeros})$ $-x^1(\text{sum of products, each from three zeros})$ $+x^0(\text{product of four zeros})$ |

The above description column also applies to the other two factorizations.

| $F(x) = (x-\omega_1)(x-\omega_2)(x-\alpha+j\beta)(x-\alpha-j\beta)$ | $F(x) = (x-\alpha_1+j\beta_1)(x-\alpha_1-j\beta_1)(x-\alpha_2+j\beta_2)(x-\alpha_2-j\beta_2)$ |
|---|---|
| $F(x) = x^4$ $-x^3(\omega_1+\omega_2+\alpha+\alpha)$ $+x^2\left(\omega_1\omega_2+\omega_1\alpha+\omega_1\alpha+\omega_2\alpha+\omega_2\alpha+\left[\alpha^2+\beta^2\right]\right)$ $-x^1\left(\omega_1\omega_2\alpha+\omega_1\omega_2\alpha+\omega_1\left[\alpha^2+\beta^2\right]+\omega_2\left[\alpha^2+\beta^2\right]\right)$ $+x^0\left(\omega_1\omega_2\left[\alpha^2+\beta^2\right]\right)$ | $F(x) = x^4$ $-x^3(\alpha_1+\alpha_1+\alpha_2+\alpha_2)$ $+x^2\left(4\alpha_1\alpha_2+\left[\alpha_1^2+\beta_1^2\right]+\left[\alpha_2^2+\beta_2^2\right]\right)$ $-x^1\left(2\alpha_1\left[\alpha_2^2+\beta_2^2\right]+2\alpha_2\left[\alpha_1^2+\beta_1^2\right]\right)$ $+x^0\left(\left[\alpha_1^2+\beta_1^2\right]\left[\alpha_2^2+\beta_2^2\right]\right)$ |

For the general polynomial $F(x) = x^n + k_{n-1}x^{n-1} + k_{n-2}x^{n-2} + \bullet\bullet\bullet + k_1x^1 + k_0$, the following table summarizes the relationships between its coefficients and its zeros.

| |
|---|
| $k_{n-1} = -(\text{sum of the zeros})$ |
| $k_{n-2} = +(\text{sum of products of zeros taken two at a time})$ |
| $k_{n-3} = -(\text{sum of products of zeros taken three at a time})$ |
| $\bullet$ |
| $\bullet$ |
| $k_1 = (-1)^{n-1} * (\text{sum of products of zeros taken } (n-1) \text{ at a time})$ |
| $k_0 = (-1)^n * (\text{product of all the zeros})$ |

# Polynomials – Programming in Chapter 4

One of the programs used in this chapter is **Sub descartes**. The polynomial is

$$F(x) = C_n x^n + C_{n-1} x^{n-1} + C_{n-2} x^{n-2} + \cdots + C_1 x + C_0.$$

The program • forms $F(-x) = (-1)^n C_n x^n + (-1)^{n-1} C_{n-1} x^{n-1} + (-1)^{n-2} C_{n-2} x^{n-2} + \cdots + (-1) C_1 x + C_0$;

• from F(x) and F(-x), removes the terms with zero coefficients;
• for F(x), counts the number of sign changes in its coefficients;
• for F(-x), counts the number of sign changes in its coefficients.

The following is a listing of **Sub descartes**.

| | |
|---|---|
| ```Sub descartes()```<br>```Dim C(20), Cm(20), rC(20), rCm(20)```<br>```N = Cells(2, 1)``` | • Read the order of F(x). |
| ```j = 1```<br>```For i = N To 0 Step -1```<br>```  C(i) = Cells(4, j)``` | • Read the coefficients of F(x). |
| ```  Cm(i) = C(i) * (-1) ^ i```<br>```  Cells(6, j) = Cm(i)``` | • Form F(-x). |
| ```  j = j + 1```<br>```Next i```<br>```j = 0```<br>```For i = 0 To N```<br>```  If C(i) <> 0 Then```<br>```    rC(j) = C(i)``` | |
| ```    rCm(j) = Cm(i)```<br>```    j = j + 1```<br>```  End If```<br>```Next i```<br>```reducedN = j - 1``` | • From F(x) and F(-x), remove terms with zero coefficients. |
| ```Count = 0```<br>```For i = 1 To reducedN```<br>```  If (rC(i) * rC(i - 1)) < 0 Then Count = Count + 1```<br>```Next i```<br>```Cells(8, 1) = Count``` | • For F(x), count the sign changes in its coefficients. |
| ```Count = 0```<br>```For i = 1 To reducedN```<br>```  If (rCm(i) * rCm(i - 1)) < 0 Then Count = Count + 1```<br>```Next i```<br>```Cells(9, 1) = Count``` | • For F(-x), count the sign changes in its coefficients. |
| ```Cells(3, 1) = "    F( x)"```<br>```Cells(5, 1) = "    F(-x)"```<br>```Cells(8, 2) = " = number of sign changes in F(x)"```<br>```Cells(9, 2) = " = number of sign changes in F(-x)"```<br>```End Sub``` | • Print labels. |

The following is the spreadsheet for program **Sub descartes**.

| Program Sub descartes Applied to Example Polynomial 1, $F(x) = x^4 + x^3 + 2x^2 + 2x + 5 = 0$ | |
| --- | --- |
| Input | The order of the polynomial and the coefficients in descending order<br><br><table><tr><td></td><td>4</td><td></td><td></td><td></td></tr><tr><td>F( x )</td><td></td><td></td><td></td><td></td></tr><tr><td></td><td>1</td><td>1</td><td>2</td><td>2</td><td>5</td></tr></table> |
| Output | <table><tr><td>F(-x)</td><td></td><td></td><td></td><td></td></tr><tr><td></td><td>1</td><td>-1</td><td>2</td><td>-2</td><td>5</td></tr></table><br><br>0 = number of sign changes in F(x)<br>4 = number of sign changes in F(-x) |
| Conclusions | There can be no real zeroes in the right-half plane,<br>and<br>there can be either 4 , 2, or 0 real zeroes in the left-half plane.<br><br>Note: Recall that Descartes' rule says nothing directly about complex zeroes. |
| Actual Zeroes | $F(x) = \underbrace{(x - .58 + j1.35)(x - .58 - j1.35)}_{\text{RHP Complex}}\underbrace{(x + 1.08 + j1.07)(x + 1.08 - j1.07)}_{\text{LHP Complex}}$ |

Another program used in this chapter is **Sub routh_array**. The polynomial is

$$F(x) = x^n + k_2 x^{n-1} + k_3 x^{n-2} + \cdots + k_n x + k_{n+1}.$$

The program builds the Routh array.

|  | $j = 1$ | $j = 2$ | $j = 3$ | $j = 4$ | $\bullet\bullet\bullet$ |
|---|---|---|---|---|---|
| $i = 1$ | $a_{11} = 1$ | $a_{12} = k_3$ | $a_{13} = k_5$ | $a_{14} = k_7$ | $\bullet\bullet\bullet$ |
| $i = 2$ | $a_{21} = k_2$ | $a_{22} = k_4$ | $a_{23} = k_6$ | $a_{24} = k_8$ | $\bullet\bullet\bullet$ |
| $\bullet$ $\bullet$ $\bullet$ | $i$ goes from 3 to $(N+1)$. For each $i$, $j = 1$ to $X^\Delta$ : $$a_{i,j} = a_{i-2,j+1} - a_{i-1,j+1}\frac{a_{i-2,1}}{a_{i-1,1}}$$ $^\Delta$X varies with row. A row ends when the rest of the row becomes all zeros. | | | | |

The following is a listing of **Sub routh_array**.

| | |
|---|---|
| ```
Sub routh_array()
 Dim a(20, 20)
 N = Cells(2, 1)
  j = 1
 For i = 1 To N + 1 Step 2
   a(1, j) = Cells(4, i)
   a(2, j) = Cells(4, i + 1)
   j = j + 1
 Next i

 For i = 3 To N + 1
  If a(i - 1, 1) > -0.0001 And a(i - 1, 1) < 0 Then a(i - 1, 1) = -0.0001
  If a(i - 1, 1) >= 0 And a(i - 1, 1) < 0.0001 Then a(i - 1, 1) = 0.0001
  For j = 1 To N
    a(i, j) = a(i - 2, j + 1) - a(i - 1, j + 1) * a(i - 2, 1) / a(i - 1, 1)
  Next j
 Next i

 For i = 1 To N + 1
   For j = 1 To (0.5 * N + 1)
     Cells(i, j + N + 6) = a(i, j)
   Next j
 Next i
End Sub
``` | • Read the polynomial coefficients. <br> • Form the first two rows of the array. <br><br><br> • This prevents division-by-zero. <br><br><br> • Compute $a_{i,j} = a_{i-2,j+1} - a_{i-1,j+1}\frac{a_{i-2,1}}{a_{i-1,1}}$. <br><br><br><br> • Print. |
```

The following is the spreadsheet for program **Sub routh_array**.

| Program Sub routh Applied to Example Polynomial 1, $F(x)=x^4+x^3+2x^2+2x+5=0$ | |
|---|---|
| **Input** | The order of the polynomial and the coefficients in descending order<br><br>     4<br>F( x )<br>  1    1    2    2    5 |
| **Output** | The Routh Array<br><br>1   2   5<br>1   2<br>0.0001  5  0<br>-49998  0  0<br>5  0  0 |
| **Conclusions** | There are two zeroes in the right-half plane,<br>and<br>there are two zeroes in the left-half plane.<br><br>Note: Recall that the Routh Array does not distinguish between real and complex zeroes. |
| **Actual Zeroes** | $F(x)=\underbrace{(x-.58+j1.35)(x-.58-j1.35)}_{\text{RHP Complex}}\underbrace{(x+1.08+j1.07)(x+1.08-j1.07)}_{\text{LHP Complex}}$ |

# Chapter 5: Finding All the Zeros of a Polynomial

In chapter 4, we derived the following equation for a spring-mass-damper system.

$$m\frac{\Delta V}{\Delta t} + d\frac{\Delta Z}{\Delta t} + kZ = 0 \qquad \text{Equation one}$$

We transformed this equation.

$$\left(m\alpha^2 + d\alpha + k\right)Z = 0$$

In chapter 3, we showed that for non-zero Z,

$$\left(m\alpha^2 + d\alpha + k\right) = 0 \qquad \text{Equation two}$$

The α's that solve this polynomial are called zeros. Solving for zeros is the subject of this chapter. Let's put **Equation one** into matrix form.

$$\frac{\Delta V}{\Delta t} = -\frac{d}{m}\frac{\Delta Z}{\Delta t} - \frac{k}{m}Z \quad \text{and} \quad \frac{\Delta Z}{\Delta t} = V$$

Rewriting these equations we get

$$\frac{\Delta V}{\Delta t} = -\frac{d}{m}V - \frac{k}{m}Z$$

$$\frac{\Delta Z}{\Delta t} = V$$

From chapter 4, $\dfrac{\Delta V}{\Delta t} = \alpha V$ and $\dfrac{\Delta Z}{\Delta t} = \alpha Z$. Therefore,

$$\alpha V = -\frac{d}{m}V - \frac{k}{m}Z$$

$$\alpha Z = V$$

In matrix form,
$$\begin{bmatrix} \alpha & 0 \\ 0 & \alpha \end{bmatrix}\begin{bmatrix} V \\ Z \end{bmatrix} = \begin{bmatrix} -\dfrac{d}{m} & -\dfrac{k}{m} \\ 1 & 0 \end{bmatrix}\begin{bmatrix} V \\ Z \end{bmatrix} \qquad \text{Equation three}$$

Rewriting again,
$$\begin{bmatrix} \left(\alpha + \dfrac{d}{m}\right) & \dfrac{k}{m} \\ -1 & \alpha \end{bmatrix}\begin{bmatrix} V \\ Z \end{bmatrix} = 0$$

In chapter 3, we showed that non-zero solutions exist only if

$$\begin{vmatrix} \left(\alpha + \dfrac{d}{m}\right) & \dfrac{k}{m} \\ -1 & \alpha \end{vmatrix} = 0 \qquad \text{Equation four}$$

Equation four is the matrix version of the polynomial Equation two. For equations higher than second order, Equation four is much easier to solve. The following is a discussion on how to solve Equation four.

Let's introduce some notation into Equation three.

$$\bullet \quad \begin{bmatrix} \alpha & 0 \\ 0 & \alpha \end{bmatrix} = \alpha \begin{bmatrix} 1 & 0 \\ 0 & 1 \end{bmatrix} = \alpha I, \text{ where } I \text{ is the identity matrix}$$

$$\bullet \quad \begin{bmatrix} -\dfrac{d}{m} & -\dfrac{k}{m} \\ 1 & 0 \end{bmatrix} = A$$

$$\bullet \quad \begin{bmatrix} V \\ Z \end{bmatrix} = x$$

Equation three can now be written as $\qquad \alpha Ix = Ax \qquad$ Equation five

This is a very famous algebraic equation. It has its own name: *The Eigenvalue Problem*.

Rewriting Equation five we get $\qquad (A - \alpha I)x = 0$.

To solve for non-zero x, $\qquad |A - \alpha I| = 0$.

Assume for the moment that A is an upper-triangular matrix.

$$A = \begin{bmatrix} a_{11} & a_{12} & a_{13} & \bullet & a_{1n} \\ 0 & a_{22} & a_{23} & \bullet & a_{2n} \\ 0 & 0 & a_{33} & \bullet & a_{3n} \\ \bullet & \bullet & \bullet & \bullet & \bullet \\ 0 & 0 & 0 & \bullet & a_{nn} \end{bmatrix}.$$

Then $\qquad |A - \alpha I| = \begin{vmatrix} a_{11}-\alpha & a_{12} & a_{13} & \bullet & a_{1n} \\ 0 & a_{22}-\alpha & a_{23} & \bullet & a_{2n} \\ 0 & 0 & a_{33}-\alpha & \bullet & a_{3n} \\ \bullet & \bullet & \bullet & \bullet & \bullet \\ 0 & 0 & 0 & \bullet & a_{nn}-\alpha \end{vmatrix}.$

From chapter 3, this becomes: $|A - \alpha I| = (\alpha - a_{11})(\alpha - a_{22})(\alpha - a_{33}) \bullet \bullet \bullet (\alpha - a_{nn})$. We see that if we could change a polynomial into an upper-triangular matrix, we would have its zeros. Let's first talk about changing a polynomial into a general matrix.

Suppose we have the following: $\alpha^3 + c_2\alpha^2 + c_1\alpha + c_0 = 0$. This polynomial comes from the equation

$$\alpha^3 z + c_2\alpha^2 z + c_1\alpha z + c_0 z = 0.$$

Let's let $y_1 = z$, $y_2 = \alpha z$ and $y_3 = \alpha^2 z$. We can then write the following series of equations

$$\alpha y_1 = \alpha z = y_2$$
$$\alpha y_2 = \alpha^2 z = y_3$$
$$\alpha y_3 = \alpha^3 z = -c_0 z - c_1\alpha z - c_2\alpha^2 z = -c_0 y_1 - c_1 y_2 - c_2 y_3$$

Rewriting,

$$\alpha y_1 = y_2$$
$$\alpha y_2 = y_3$$
$$\alpha y_3 = -c_0 y_1 - c_1 y_2 - c_2 y_3$$

In matrix form,

$$\alpha \begin{bmatrix} 1 & 0 & 0 \\ 0 & 1 & 0 \\ 0 & 0 & 1 \end{bmatrix} \begin{bmatrix} y_1 \\ y_2 \\ y_3 \end{bmatrix} = \begin{bmatrix} 0 & 1 & 0 \\ 0 & 0 & 1 \\ -c_0 & -c_1 & -c_2 \end{bmatrix} \begin{bmatrix} y_1 \\ y_2 \\ y_3 \end{bmatrix}$$

Finally,

$$\alpha I \begin{bmatrix} y_1 \\ y_2 \\ y_3 \end{bmatrix} = A \begin{bmatrix} y_1 \\ y_2 \\ y_3 \end{bmatrix}$$

For the general polynomial $\alpha^n + k_{n-1}\alpha^{n-1} + k_{n-2}\alpha^{n-2} + \bullet\bullet\bullet + k_3\alpha^3 + k_2\alpha^2 + k_1\alpha + k_0 = 0$, the same procedure yields the following matrix.

$$\begin{bmatrix} \alpha y_1 \\ \alpha y_2 \\ \alpha y_3 \\ \bullet \\ \alpha y_{n-1} \\ \alpha y_n \end{bmatrix} = \begin{bmatrix} 0 & 1 & 0 & 0 & \bullet & 0 & 0 \\ 0 & 0 & 1 & 0 & \bullet & 0 & 0 \\ 0 & 0 & 0 & 1 & \bullet & 0 & 0 \\ \bullet & \bullet & \bullet & \bullet & \bullet & \bullet & \bullet \\ 0 & 0 & 0 & 0 & \bullet & 0 & 1 \\ -k_0 & -k_1 & -k_2 & -k_3 & \bullet & -k_{n-2} & -k_{n-1} \end{bmatrix} \begin{bmatrix} y_1 \\ y_2 \\ y_3 \\ \bullet \\ y_{n-1} \\ y_n \end{bmatrix}$$

Having changed a polynomial into a general matrix, we turn to the problem of solving $|A - \alpha I| = 0$. Wilkinson[3] shows the QR method. This method changes a general matrix into an upper-triangular matrix. It uses linear algebra, which is not covered in this book. However, the derivation with references

---

[3] Wilkinson, J., *The Algebraic Eigenvalue Problem* (Oxford, UK: Oxford University Press, 1965).

is in appendix B. In this chapter, we will demonstrate how the QR method works and present a program that uses it.

Suppose we want the zeros of $F(\alpha) = \alpha^3 + 2.5\alpha^2 - 11.5\alpha + 5$. First we change it into a matrix. $F(\alpha)$ comes from the equation $\alpha^3 z + 2.5\alpha^2 z - 11.5\alpha z + 5z = 0$. Let's let $y_1 = z$, $y_2 = \alpha z$ and $y_3 = \alpha^2 z$. Now we write the following equations.

$$\alpha y_1 = \alpha z = y_2$$

$$\alpha y_2 = \alpha^2 z = y_3$$

$$\alpha y_3 = \alpha^3 z = -5z + 11.5\alpha z - 2.5\alpha^2 z = -5y_1 + 11.5y_2 - 2.5y_3$$

In matrix form we get
$$\begin{bmatrix} \alpha & 0 & 0 \\ 0 & \alpha & 0 \\ 0 & 0 & \alpha \end{bmatrix}\begin{bmatrix} y_1 \\ y_2 \\ y_3 \end{bmatrix} = \begin{bmatrix} 0 & 1 & 0 \\ 0 & 0 & 1 \\ -5 & 11.5 & -2.5 \end{bmatrix}\begin{bmatrix} y_1 \\ y_2 \\ y_3 \end{bmatrix}$$

The QR method iteratively transforms $\begin{bmatrix} 0 & 1 & 0 \\ 0 & 0 & 1 \\ -5 & 11.5 & -2.5 \end{bmatrix}$ into upper-triangular form as shown in the following table.

| Iteration | The Matrix | | | Iteration | The Matrix | | |
|---|---|---|---|---|---|---|---|
| start | 0 | 1 | 0 | 5 | -4.91 | 8.78 | 7.63 |
| | 0 | 0 | 1 | | 0.07 | 1.92 | 1.46 |
| | -5 | 11.5 | -2.5 | | 0 | -0.01 | 0.49 |
| 1 | -2.5 | 5 | -11.5 | 6 | -5.04 | 8.65 | -7.64 |
| | 0 | 0 | 1 | | -0.03 | 2.04 | -1.58 |
| | -1 | 0 | 0 | | 0 | 0 | 0.5 |
| 2 | -6.47 | 8.91 | -4.64 | 7 | -4.99 | 8.69 | 7.65 |
| | -1.59 | 3.97 | -1.86 | | 0.01 | 1.99 | 1.53 |
| | -0.37 | 0.93 | 0 | | 0 | 0 | 0.5 |
| 3 | -4.45 | 9.64 | 7.01 | 8 | -5.01 | 8.67 | -7.65 |
| | 0.39 | 1.54 | 1.01 | | 0 | 2.01 | -1.55 |
| | -0.02 | -0.09 | 0.41 | | 0 | 0 | 0.5 |
| 4 | -5.23 | 8.61 | -7.46 | 9 | -5 | 8.68 | 7.65 |
| | -0.2 | 2.26 | -1.73 | | 0 | 2 | 1.55 |
| | 0 | 0.03 | 0.47 | | 0 | 0 | 0.5 |

The diagonal of the last matrix shows that the zeros of the polynomial are 0.5, 2 and $-5$. To check this, substitute $\alpha = 0.5$ into F($\alpha$). $F(0.5) = 0.5^3 + 2.5*0.5^2 - 11.5*0.5 + 5 = 0$. The same is true for $\alpha = 2$ and $-5$. From chapter 4, we get two additional ways to check these zeros.

- The sum of the zeros is $-2.5$ which equals the term $(-k_{n-1} = -2.5)$.

- The product of the zeros is $-5$ which equals the term $\left((-1)^3 k_0 = -5\right)$.

We will use these three items to check the accuracy of the zeros computed by the QR method.

In chapter 1, we discussed how to compute the sum and product of complex zeros. To compute the value of a polynomial requires that complex zeros be raised to integer powers. The following is a discussion on how to do this.

### Computing the Value of a Polynomial at a Complex Zero

Define a complex zero as $\alpha = x + jy$. The following figure illustrates how to convert this to polar coordinates.

$$r = \sqrt{x^2 + y^2}$$

$$\theta = \tan^{-1} \frac{y}{x}$$

$$\alpha = r\cos\theta + j\,r\sin\theta$$

$$\alpha^2 = r^2(\cos\theta + j\sin\theta)^2 = r^2\left[(\cos\theta\cos\theta - \sin\theta\sin\theta) + j(\sin\theta\cos\theta + \cos\theta\sin\theta)\right]$$

In chapter 6 on trigonometry, we will derive that

$$\cos 2\theta = \cos\theta\cos\theta - \sin\theta\sin\theta \text{ and } \sin 2\theta = \sin\theta\cos\theta + \cos\theta\sin\theta.$$

Therefore, $\alpha^2 = r^2(\cos 2\theta + j\sin 2\theta)$. Similarly, $\alpha^3 = r^3(\cos 3\theta + j\sin 3\theta)$. This leads to DeMoivre's theorem, which is $\alpha^k = r^k(\cos k\theta + j\sin k\theta)$.

The polynomial to be evaluated is $F(\alpha) = c_{n+1}\alpha^n + c_n\alpha^{n-1} + \cdots + c_2\alpha + c_1$, where $c_{n+1} = 1$. Substituting $\alpha^k = r^k(\cos k\theta + j\sin k\theta)$ and $\alpha^0 = 1$ into $F(\alpha)$ yields

$$F(\alpha) = c_1 + \sum_{k=1}^{n} c_{k+1} r^k \cos k\theta + j\sum_{k=1}^{n} c_{k+1} r^k \sin k\theta = \text{Re} + j\,\text{Imag}$$

$$\text{Amp} = \sqrt{\text{Re}^2 + \text{Imag}^2}$$

If $\alpha$ is a zero of F($\alpha$), the value of Amp will be zero. But because of roundoff, there will be tolerances. For large n, Amp can only be expected to be close to zero. Note that Amp is computed for each zero.

## The Program That Computes All the Zeros of a Polynomial

The following is a block diagram of this program.

```
Sub zeros_main()

    • Read in the polynomial.
    • Call Sub eig to compute the zeros.
    • Print out the zeros.
    • Print out the checkout results.
            • Evaluate the polynomial at each zero.
            • Compute the sum of the zeros.
            • Compute the product of the zeros.

End Sub
```

```
Sub eig

QR iterations with calls to Sub Q_R, Sub hess,
Sub deflate and Sub compute_guess.

End Sub
```

The following table contains a listing of **Sub zeros_main**. See the note at the bottom.

| Code | Comments |
|---|---|
| Option Base 1<br>Sub zeros_main()<br>Dim A(30, 30), WR(30), WI(30), c(30), Cin(30)<br>N = Cells(5, 1)<br>For i = 1 To N<br>  For j = 1 To N<br>    A(i, j) = 0<br>  Next j<br>Next i<br>For i = 1 To N + 1<br>  Cin(N + 2 - i) = Cells(6, i)<br>Next i<br>For i = 1 To N + 1<br>  c(i) = Cin(i) / Cin(N + 1)<br>Next i<br>For i = 1 To N - 1<br>  A(i, i + 1) = 1<br>Next i<br>For i = 1 To N<br>  A(N, i) = -c(i)<br>Next i<br>Call eig(N, A, WR, WI)<br>For i = 1 To N<br>  Cells(8 + i, 1) = Application.Round(WR(i), 3)<br>  Cells(8 + i, 2) = Application.Round(WI(i), 3)<br>Next i<br>eigsum = 0: prod = Application.Complex(1, 0): psign = 1<br>For i = 1 To N<br>  eigsum = eigsum + WR(i)<br>  Root = Application.Complex(WR(i), WI(i))<br>  prod = Application.ImProduct(prod, Root)<br>  prodR = Application.ImAbs(prod): psign = psign * WR(i)<br>Next i<br>  If psign < 0 Then prodR = -prodR<br>For i = 1 To N<br>  r = Sqr(WR(i) ^ 2 + WI(i) ^ 2)<br>  If WR(i) = 0 Then<br>    thetar = 90 / 57.296<br>  Else<br>    thetar = Application.Atan2(WR(i), WI(i))<br>  End If<br>  Re = c(1): Im = 0<br>  For K = 1 To N<br>    delRe = c(K + 1) * r ^ K * Cos(K * thetar)<br>    delIm = c(K + 1) * r ^ K * Sin(K * thetar)<br>    Re = Re + delRe: Im = Im + delIm<br>  Next K<br>  Amp = Sqr(Re ^ 2 + Im ^ 2)<br>  Cells(i + 8, 4) = Application.Round(Amp, 2)<br>Next i<br>Cells(N + 11, 1) = Application.Round(eigsum, 3)<br>Cells(N + 11, 2) = -c(N): Cells(N + 11, 5) = c(1) * (-1) ^ N<br>Cells(N + 11, 4) = Application.Round(prodR, 3)<br>Cells(8, 1) = " WR(i)": Cells(8, 2) = " WI(i)": Cells(8, 4) = " Amp"<br>Cells(N + 10, 1) = " eigsum": Cells(N + 10, 2) = " -c(N)"<br>Cells(N + 10, 4) = " prodR": Cells(N + 10, 5) = " -1^N*c1"<br>End Sub | • Read N, the order of the polynomial.<br><br><br>• Read the coefficients of the input polynomial.<br>$$Cin_{n+1}\alpha^n + Cin_n\alpha^{n-1} + \cdots + Cin_2\alpha + Cin_1$$<br><br>• Normalize the polynomial.<br>$$c_{n+1}\alpha^n + c_n\alpha^{n-1} + \cdots + c_2\alpha + c_1$$<br><br><br>• Form the A matrix.<br><br>• Call Sub eig to compute the zeros.<br><br>• Print out the zeros: $\alpha_i = WR_i + j\,WI_i$<br><br><br>• Compute and print out for checkout:<br><br>  • Magnitude of the polynomial at each zero (Amp).<br>  • The sum of the zeros (eigsum).<br>  • The value of –c(N).<br>  • The product of the zeros (prodR).<br>  • The value of $(-1)^N$ times c(1) |

**Note:** Sub zeros_main, Sub eig, Sub hess, Sub compute_guess, Sub deflate, and Sub Q_R should be in the same code window. Listings are in appendix B.

The following table shows the spreadsheet of the input and the output of the program Sub zeros_main.

| The spreadsheet of Sub zeros_main to find the zeros of $F(x) = x^3 + 2.5x^2 - 11.5x + 5$ | | | | | |
|---|---|---|---|---|---|
| 3 | | | | | |
| 1 | 2.5 | -11.5 | | 5 | |
| | | | | | |
| WR(i) | WI(i) | | | Amp | |
| 2 | 0 | | | 0 | |
| 0.5 | 0 | | | 0 | |
| -5 | 0 | | | 0 | |
| | | | | | |
| eigsum | -c(N) | | | prodR | -1^N*c1 |
| -2.5 | -2.5 | | | -5 | -5 |

- The zeros (WR(i) + $j$ WI(i)) are at: 2, 0.5 and -5.
- For checkout: Amp = 0 for each zero; eigsum = -c(N); prodR = $(-1)^N c(1)$.

The following are five examples.

## Example Z.1: Yaw-Roll Airplane Motion

In reference[4], the QR method was used to compute the eigenvalues of the following matrix.

$$\begin{bmatrix} -0.1276 & -0.9927 & 0.0731 & 0 \\ 1.6402 & -0.3384 & 0.0171 & -0.0405 \\ 0 & 0 & 0 & 1 \\ -2.0922 & 0.4904 & -2.2024 & -4.2629 \end{bmatrix}$$

Its eigenvalues are $(x+0.57)(x+0.24+j1.295)(x+0.24-j1.295)(x+3.696)$. These eigenvalues were multiplied together, producing the polynomial $x^4 + 4.746x^3 + 5.868x^2 + 8.372x + 3.654$. This polynomial was then input to **Sub zeros_main** as shown in the following spreadsheet.

| | | | | |
|---|---|---|---|---|
| 4 | | | | |
| 1 | 4.746 | 5.868 | 8.372 | 3.654 |
| | | | | |
| WR(i) | WI(i) | | Amp | |
| -0.573 | 0 | | 0 | |
| -0.237 | 1.292 | | 0 | |
| -0.237 | -1.292 | | 0 | |
| -3.699 | 0 | | 0 | |
| | | | | |
| eigsum | -c(N) | | prodR | -1^N*c1 |
| -4.746 | -4.746 | | 3.654 | 3.654 |

The spreadsheet shows the computed zeros and the results from the checkout parameters. The zeros are compared to the input eigenvalues in the adjacent plot. The comparison is excellent.

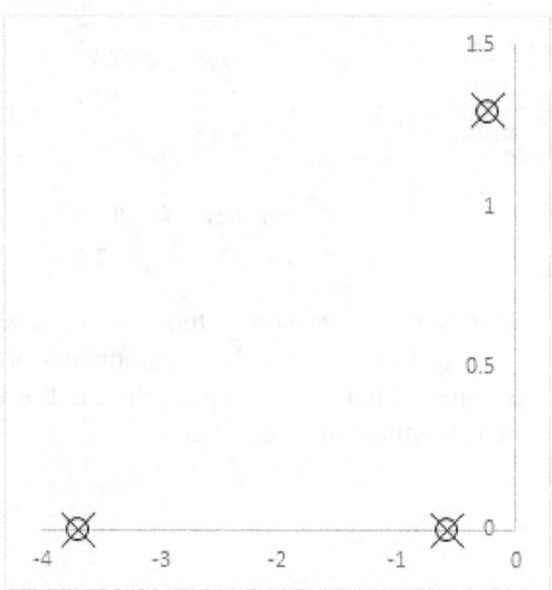

[4] Hauser, F., *Eigenvalues and Eigenvectors*, (North Charleston, SC: Createspace Independent Publishing, 2016).

## Example Z.2: Double Inverted Pendulum

As with example Z.1, the QR method was used to compute the eigenvalues of the following matrix.

$$\begin{bmatrix} 0 & 0 & 1 & 0 & 0 & 0 \\ 0 & 0 & 0 & 1 & 0 & 0 \\ -80.2 & 89.5 & -5.26 & 9.2 & 0.126 & 0.36 \\ -22.9 & 20.7 & 1.25 & -2.2 & -0.03 & -0.087 \\ 0 & 0 & 0 & 0 & 0 & 1 \\ 50.8 & -44.6 & 2.04 & -3.6 & -0.05 & -0.14 \end{bmatrix}$$

Its eigenvalues are

$$(x+0.2+j0.35)(x+0.2-j0.35)(x+1.18+j2.75)(x+1.18-j2.75)(x+2.42+j5.51)(x+2.42-j5.51).$$

These eigenvalues were multiplied together, producing the polynomial

$$x^6 + 7.6x^5 + 59.6363x^4 + 152.6202x^3 + 385.0369x^2 + 150.6597x + 52.7012.$$

This polynomial was then input to Sub **zeros_main** as shown in the following spreadsheet.

| 6 | | | | | | |
|---|---|---|---|---|---|---|
| 1 | 7.6 | 59.6363 | 152.6202 | 385.0369 | 150.6597 | 52.7012 |
| | | | | | | |
| WR(i) | WI(i) | | Amp | | | |
| -0.2 | 0.35 | | 0 | | | |
| -0.2 | -0.35 | | 0 | | | |
| -1.18 | 2.75 | | 0 | | | |
| -1.18 | -2.75 | | 0 | | | |
| -2.42 | 5.51 | | 0.05 | | | |
| -2.42 | -5.51 | | 0.05 | | | |
| | | | | | | |
| eigsum | -c(N) | | prodR | -1^N*c1 | | |
| -7.6 | -7.6 | | 52.701 | 52.7012 | | |

The spreadsheet shows the computed zeros and the results from the checkout parameters. The zeros are compared to the input eigenvalues in the adjacent plot. The comparison is excellent.

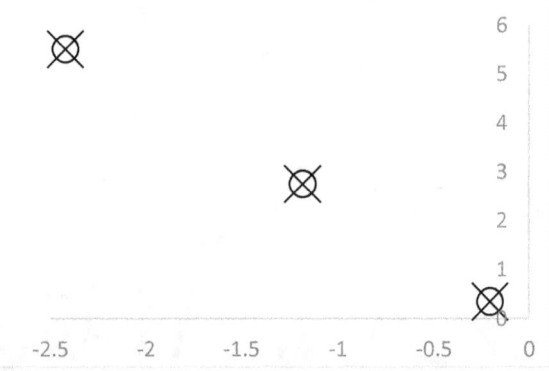

## Example Z.3: Launch Vehicle Autopilot

As with examples Z.1 and Z.2, the QR method was applied to the following matrix.

$$
\begin{bmatrix}
0 & 1 & 0 & 0 & 0 & 0 & 0 & 0 & 0 & 0 & 0 & 0 & 0 & 0 \\
3.5 & 0 & 0.00269 & 0 & 0 & 0 & 0 & 0 & 0 & -0.06 & 0 & 0 & 4.6 & 0 \\
-112.727 & 0 & -0.02797 & 0 & 0 & 0 & 0 & 0 & 0 & 0 & 0 & 0 & 61.8182 & 0 \\
0 & 0 & 0 & 0 & 1 & 0 & 0 & 0 & 0 & 0 & 0 & 0 & 0 & 0 \\
0 & 4900 & 0 & -4900 & -70 & 0 & 49 & 0 & 24.5 & 0 & 0 & 0 & 0 & 0 \\
0 & 0 & 0 & 0 & 0 & 0 & 1 & 0 & 0 & 0 & 0 & 0 & 0 & 0 \\
0 & 0 & 0 & 0 & 0 & -400 & -0.4 & 0 & 0 & 0 & 0 & 0 & -213.836 & 0 \\
0 & 0 & 0 & 0 & 0 & 0 & 0 & 0 & 1 & 0 & 0 & 0 & 0 & 0 \\
0 & 0 & 0 & 0 & 0 & 0 & 0 & -900 & -0.6 & 0 & 0 & 0 & -425 & 0 \\
0 & 0 & 0 & 0 & 0 & 0 & 0 & 0 & 0 & 0 & 1 & 0 & 0 & 0 \\
0 & 0 & 0 & 0 & 0 & 0 & 0 & 0 & 0 & -26.01 & -0.102 & 0 & -17.9 & 0 \\
0 & 0 & 0 & -2520 & 0 & 0 & 0 & 0 & 0 & 0 & 0 & -84 & -3600 & -7560 \\
0 & 0 & 0 & 0 & 0 & 0 & 0 & 0 & 0 & 0 & 0 & 1 & 0 & 0 \\
35 & 0 & 0 & 0 & 0 & 0.35 & 0 & 0.175 & 0 & 0 & 0 & 0 & 0 & -35
\end{bmatrix}
$$

It computed the eigenvalues listed in the following table. The table also lists the polynomial that resulted from multiplying the eigenvalues.

| The Eigenvalues of the Matrix | | The Polynomial (Product of the Eigenvalues) |
|---|---|---|
| WR(i) | WI(i) | $x^{14}$ |
| -0.1346 | 0 | $+190.11\,x^{13}$ |
| -1.5168 | 2.1407 | $+2.130342E+04\,x^{12}$ |
| -1.5168 | -2.1407 | $+1.439885E+06\,x^{11}$ |
| -0.0315 | 5.0992 | $+6.888428E+07\,x^{10}$ |
| -0.0315 | -5.0992 | $+2.301969E+09\,x^{9}$ |
| 0.4503 | 19.4115 | $+6.364700E+10\,x^{8}$ |
| 0.4503 | -19.4115 | $+1.349727E+12\,x^{7}$ |
| -0.0641 | 29.2568 | $+1.814778E+13\,x^{6}$ |
| -0.0641 | -29.2568 | $+2.802555E+14\,x^{5}$ |
| -35.2156 | 0 | $+1.162837E+15\,x^{4}$ |
| -41.1262 | 43.4577 | $+7.888950E+15\,x^{3}$ |
| -41.1262 | -43.4577 | $+1.898774E+16\,x^{2}$ |
| -35.1016 | 60.2154 | $+3.779537E+16\,x$ |
| -35.1016 | -60.2154 | $+4.609516E+15$ |

This polynomial was input to program **Sub zeros_main**. The following table shows:

- the input polynomial;
- the output zeros along with the checkout parameters;
- a plot comparing the zeros with the input eigenvalues.

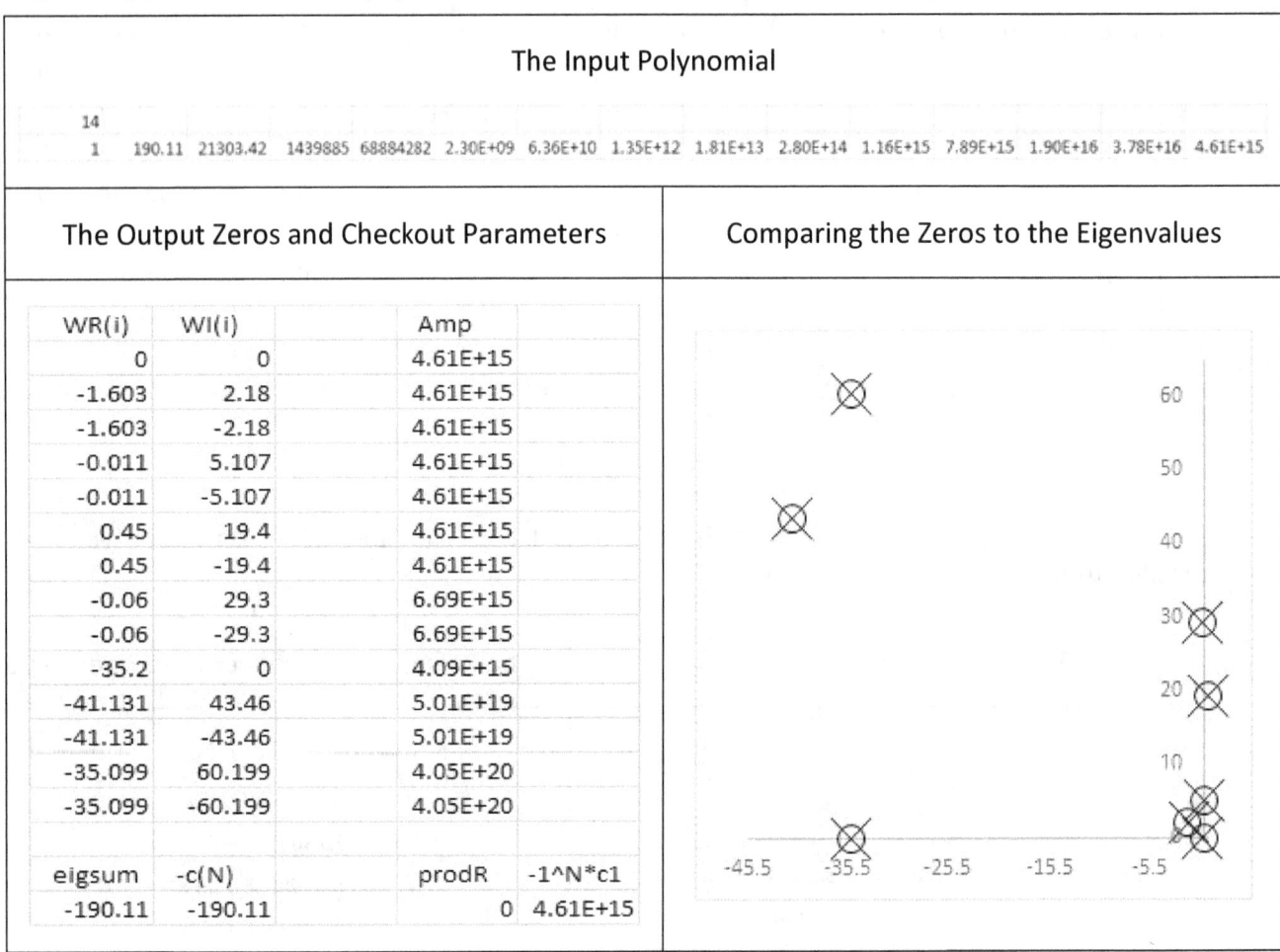

| The Input Polynomial | | | | | | | | | | | | | | | |
|---|---|---|---|---|---|---|---|---|---|---|---|---|---|---|---|
| 14 | | | | | | | | | | | | | | | |
| 1 | 190.11 | 21303.42 | 1439885 | 68884282 | 2.30E+09 | 6.36E+10 | 1.35E+12 | 1.81E+13 | 2.80E+14 | 1.16E+15 | 7.89E+15 | 1.90E+16 | 3.78E+16 | 4.61E+15 | |

**The Output Zeros and Checkout Parameters**

| WR(i) | WI(i) | | Amp | |
|---|---|---|---|---|
| 0 | 0 | | 4.61E+15 | |
| -1.603 | 2.18 | | 4.61E+15 | |
| -1.603 | -2.18 | | 4.61E+15 | |
| -0.011 | 5.107 | | 4.61E+15 | |
| -0.011 | -5.107 | | 4.61E+15 | |
| 0.45 | 19.4 | | 4.61E+15 | |
| 0.45 | -19.4 | | 4.61E+15 | |
| -0.06 | 29.3 | | 6.69E+15 | |
| -0.06 | -29.3 | | 6.69E+15 | |
| -35.2 | 0 | | 4.09E+15 | |
| -41.131 | 43.46 | | 5.01E+19 | |
| -41.131 | -43.46 | | 5.01E+19 | |
| -35.099 | 60.199 | | 4.05E+20 | |
| -35.099 | -60.199 | | 4.05E+20 | |
| | | | | |
| eigsum | -c(N) | | prodR | -1^N*c1 |
| -190.11 | -190.11 | | 0 | 4.61E+15 |

Except for the eigenvalue at −0.1346, the comparison is excellent considering the size of the polynomial.

Example Z.4: A Double Spring-Mass System

As with the first three examples, the QR method was used to compute the eigenvalues of the following matrix.

$$\begin{bmatrix} 0 & 0 & 1 & 0 \\ 0 & 0 & 0 & 1 \\ -16.667 & 6.667 & 0 & 0 \\ 10 & -10 & 0 & 0 \end{bmatrix}$$

Its eigenvalues are $(x + j4.71)(x - j4.71)(x + j2.12)(x - j2.12)$.

These eigenvalues were multiplied together, producing the polynomial $x^4 + 26.58x^2 + 99.18$.

This polynomial was then input to Sub zeros_main as shown in the following spreadsheet.

| | | | | | |
|---|---|---|---|---|---|
| 4 | | | | | |
| 1 | 0 | 26.58 | | 0 | 99.18 |

| WR(i) | WI(i) | | Amp | |
|---|---|---|---|---|
| 0 | 2.119 | | 0 | |
| 0 | -2.119 | | 0 | |
| 0 | 4.7 | | 0 | |
| 0 | -4.7 | | 0 | |

| eigsum | -c(N) | | prodR | -1^N*c1 |
|---|---|---|---|---|
| 0 | 0 | | 99.18 | 99.18 |

The spreadsheet shows the computed zeros and the results from the checkout parameters. The zeros are compared to the input eigenvalues in the adjacent plot. The comparison is excellent.

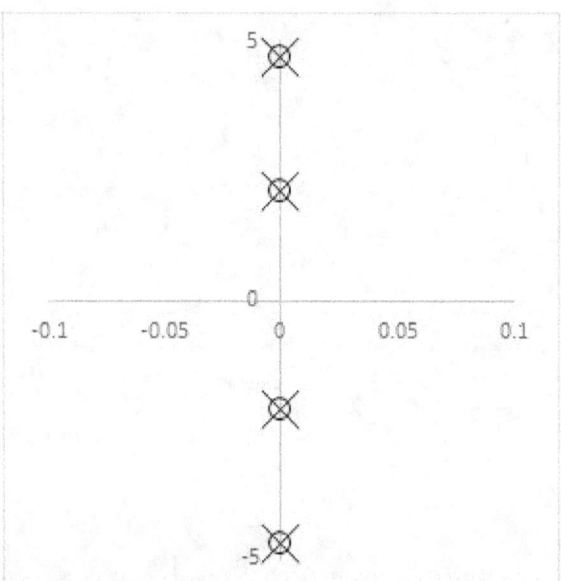

## Example Z.5: A Fourteenth-Order Butterworth Polynomial

The polynomial in this example was obtained from the zeros of a fourteenth-order Butterworth polynomial. These zeros were calculated using the formulas in Weinberg[5]. The following table lists these zeros, as well as the polynomial that results when they are multiplied together.

| The Input Zeros | | The Polynomial (Product of the Input Zeros) |
|---|---|---|
| WR(i) | WI(i) | $x^{14}$ |
| -0.11196 | 0.993712 | $+8.931404\, x^{13}$ |
| -0.11196 | -0.99371 | $+39.88499\, x^{12}$ |
| -0.99371 | 0.111965 | $+117.7337\, x^{11}$ |
| -0.99371 | -0.11197 | $+256.1214\, x^{10}$ |
| -0.33028 | 0.943883 | $+433.7284\, x^{9}$ |
| -0.33028 | -0.94388 | $+589.0206\, x^{8}$ |
| -0.53203 | 0.846724 | $+651.2664\, x^{7}$ |
| -0.53203 | -0.84672 | $+589.0206\, x^{6}$ |
| -0.94388 | 0.330279 | $+433.7284\, x^{5}$ |
| -0.94388 | -0.33028 | $+256.1214\, x^{4}$ |
| -0.84672 | 0.532032 | $+117.7337\, x^{3}$ |
| -0.84672 | -0.53203 | $+39.88499\, x^{2}$ |
| -0.70711 | 0.70711 | $+8.931404\, x$ |
| -0.70711 | -0.70711 | $+1$ |

---

[5] Weinberg, L., *Network Analysis and Synthesis*, (New York: McGraw-Hill Book Company, 1962).

This polynomial was input to program **Sub zeros_main**. The following table shows:

- the input polynomial;
- the output zeros along with the checkout parameters;
- a plot comparing the input zeros with the output zeros.

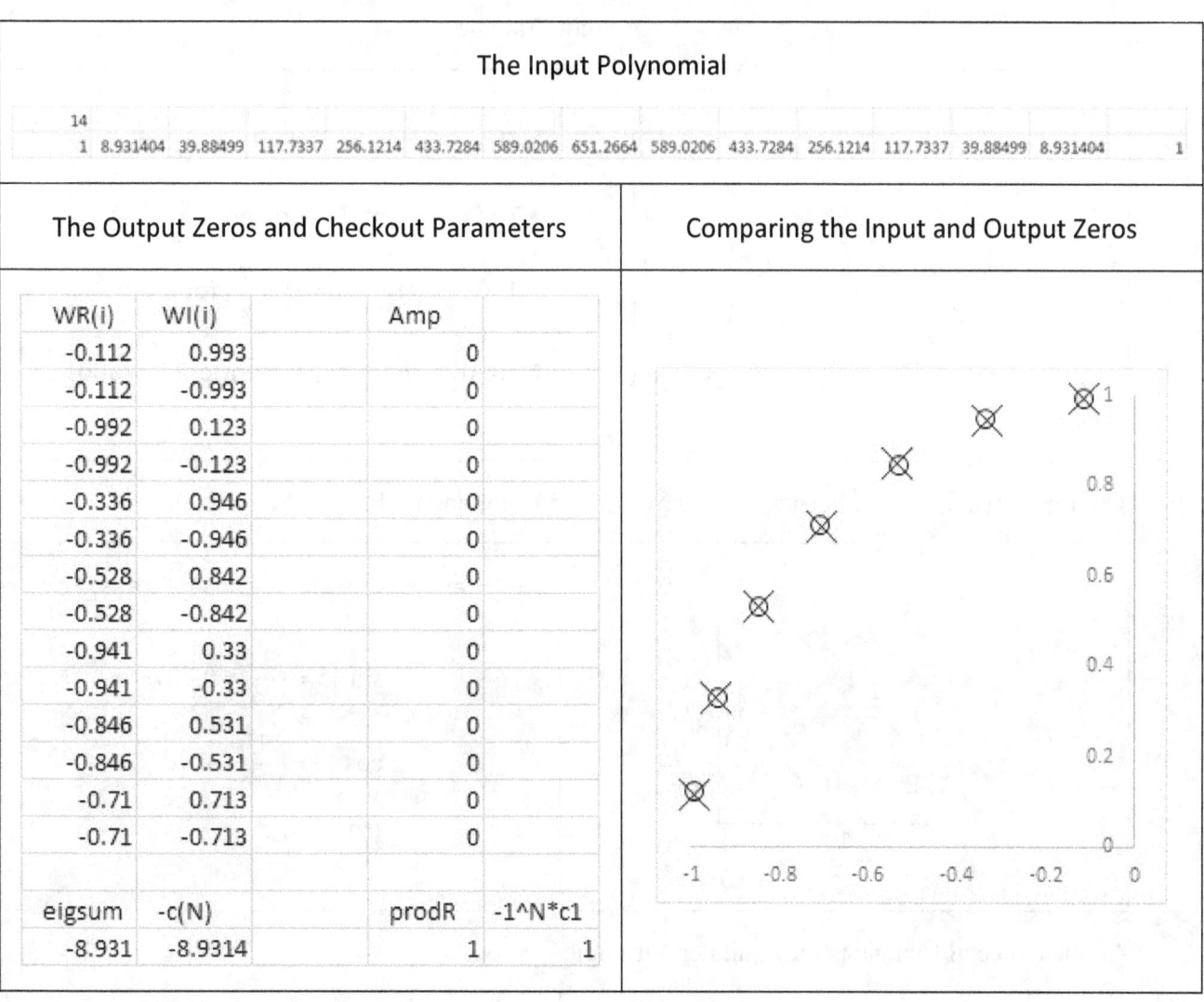

| The Input Polynomial | | | | | | | | | | | | | | |
|---|---|---|---|---|---|---|---|---|---|---|---|---|---|---|
| 14 | | | | | | | | | | | | | | |
| 1 | 8.931404 | 39.88499 | 117.7337 | 256.1214 | 433.7284 | 589.0206 | 651.2664 | 589.0206 | 433.7284 | 256.1214 | 117.7337 | 39.88499 | 8.931404 | 1 |

**The Output Zeros and Checkout Parameters** / **Comparing the Input and Output Zeros**

| WR(i) | WI(i) | | Amp | |
|---|---|---|---|---|
| -0.112 | 0.993 | | 0 | |
| -0.112 | -0.993 | | 0 | |
| -0.992 | 0.123 | | 0 | |
| -0.992 | -0.123 | | 0 | |
| -0.336 | 0.946 | | 0 | |
| -0.336 | -0.946 | | 0 | |
| -0.528 | 0.842 | | 0 | |
| -0.528 | -0.842 | | 0 | |
| -0.941 | 0.33 | | 0 | |
| -0.941 | -0.33 | | 0 | |
| -0.846 | 0.531 | | 0 | |
| -0.846 | -0.531 | | 0 | |
| -0.71 | 0.713 | | 0 | |
| -0.71 | -0.713 | | 0 | |
| | | | | |
| eigsum | -c(N) | | prodR | -1^N*c1 |
| -8.931 | -8.9314 | | 1 | 1 |

For all of the zeros, the comparison is excellent.

# Chapter 6: Trigonometry

Let's begin with the right triangle. Mathematicians have been studying this triangle for about 2500 years. Three functions based on the right triangle are defined in the following table.

| A Right Triangle | |
|---|---|
| 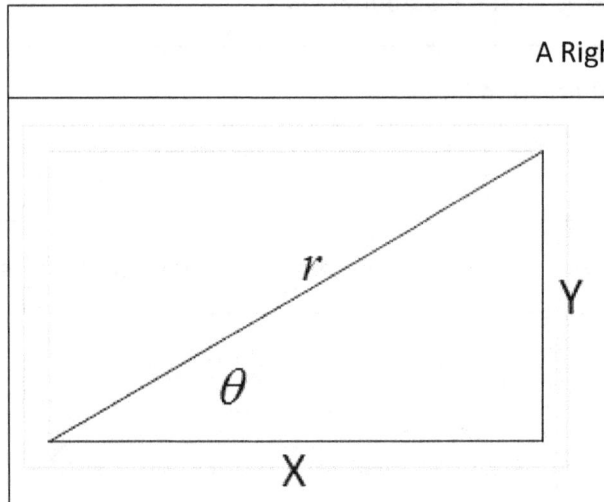 | • The sine of the angle θ is written $\sin\theta = \dfrac{Y}{r}$.<br><br>• The cosine of θ is written $\cos\theta = \dfrac{X}{r}$.<br><br>• The tangent is $\tan\theta = \dfrac{Y}{X}$ for $X \neq 0$.<br><br>Note that these are dimensionless ratios. |

The forty-five degree right triangle is shown in the following table.

| | |
|---|---|
| 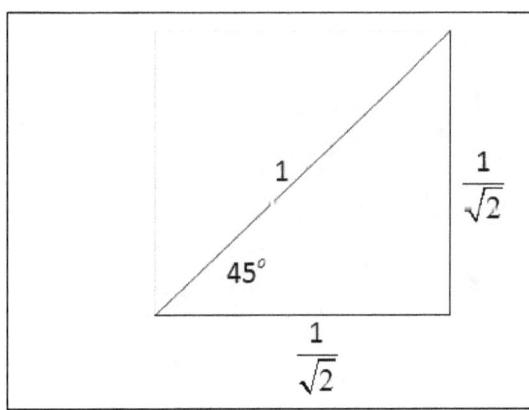 | This triangle is special because it's easy to draw.<br><br>$\sin 45 = \dfrac{1}{\sqrt{2}} = 0.707$<br><br>$\cos 45 = \dfrac{1}{\sqrt{2}} = 0.707$<br><br>$\tan 45 = 1$ |

Another special triangle is the equilateral triangle.

| | |
|---|---|
| 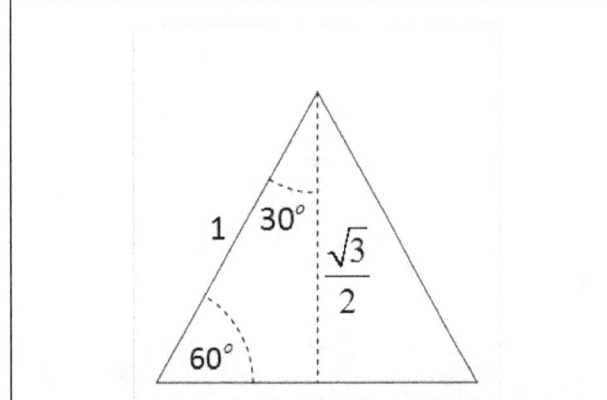 | $\sin 30 = 0.5$<br>$\cos 30 = 0.866$<br>$\tan 30 = 0.577$<br><br>and<br><br>$\sin 60 = 0.866$<br>$\cos 60 = 0.5$<br>$\tan 60 = 1.732$ |

We will focus on sine and cosine because $\tan\theta = \dfrac{\sin\theta}{\cos\theta}$. There are three others that we will not discuss because their definitions are based on sine, cosine and tangent.

- The secant of θ is $\sec\theta = \dfrac{1}{\cos\theta}$.
- The cosecant of θ is $\csc\theta = \dfrac{1}{\sin\theta}$.
- The cotangent of θ is $\cot\theta = \dfrac{1}{\tan\theta}$.

Let's define Quadrant I as having angles between 0 and 90 degrees. The following table accumulates the data we have so far. The values at 0 and 90 degrees are easily added.

| Quadrant I | |
|---|---|
| $\sin(0)=0$ <br> $\sin(30)=0.5$ <br> $\sin(45)=0.707$ <br> $\sin(60)=0.866$ <br> $\sin(90)=1$ | |
| $\cos(0)=1$ <br> $\cos(30)=0.866$ <br> $\cos(45)=0.707$ <br> $\cos(60)=0.5$ <br> $\cos(90)=0$ | |

We want to extend this table to the angles between 90 and 360 degrees. To do this, let's derive two formulas.

- The sine of the sum of two angles: $\sin(\alpha+\beta)$.
- The cosine of the sum of two angles: $\cos(\alpha+\beta)$.

These are derived in the following table. The derivation uses two right triangles that share line $\overline{OB}$. This creates the angle $(\alpha + \beta)$.

| Deriving Formulas for $\sin(\alpha + \beta)$ and $\cos(\alpha + \beta)$. | |
|---|---|
| Note: The length of line $\overline{OC}$ is unity. 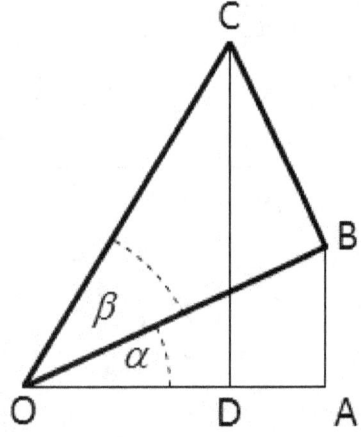 | 1. $\sin(\alpha + \beta) = \overline{CD}$ <br> 2. $\cos(\alpha + \beta) = \overline{OD} = \overline{OA} - \overline{AD}$ <br> • $\sin\beta = \overline{BC}$ <br> • $\cos\beta = \overline{OB}$ <br> • $\sin\alpha = \dfrac{\overline{AB}}{\overline{OB}}$ <br> Hence → $\overline{AB} = \overline{OB}\sin\alpha = \sin\alpha\cos\beta$ <br> • $\cos\alpha = \dfrac{\overline{OA}}{\overline{OB}}$ <br> Hence → $\overline{OA} = \overline{OB}\cos\alpha = \cos\alpha\cos\beta$ |
| Adding Line $\overline{BE}$ So That $\overline{CD} = \overline{DE} + \overline{CE}$ and $\overline{DE} = \overline{AB}$ and $\overline{AD} = \overline{BE}$ | |
| 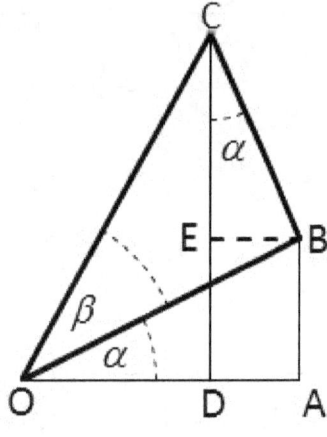 | • $\cos\alpha = \dfrac{\overline{CE}}{\overline{BC}}$ <br> Hence → $\overline{CE} = \overline{BC}\cos\alpha = \cos\alpha\sin\beta$ <br> • $\sin\alpha = \dfrac{\overline{BE}}{\overline{BC}}$ <br> Hence → $\overline{BE} = \overline{BC}\sin\alpha = \sin\alpha\sin\beta$ |
| Combining the above equations yields the following. | |

1. $\sin(\alpha + \beta) = \overline{CD} = \overline{DE} + \overline{CE} = \overline{AB} + \overline{CE} = \sin\alpha\cos\beta + \cos\alpha\sin\beta$ → $\boxed{\sin(\alpha + \beta) = \sin\alpha\cos\beta + \cos\alpha\sin\beta}$

2. $\cos(\alpha + \beta) = \overline{OD} = \overline{OA} - \overline{AD} = \overline{OA} - \overline{BE} = \cos\alpha\cos\beta - \sin\alpha\sin\beta$ → $\boxed{\cos(\alpha + \beta) = \cos\alpha\cos\beta - \sin\alpha\sin\beta}$

The following table shows the calculations that extend the sine and cosine to angles from 90 to 180 degrees. This is called Quadrant II.

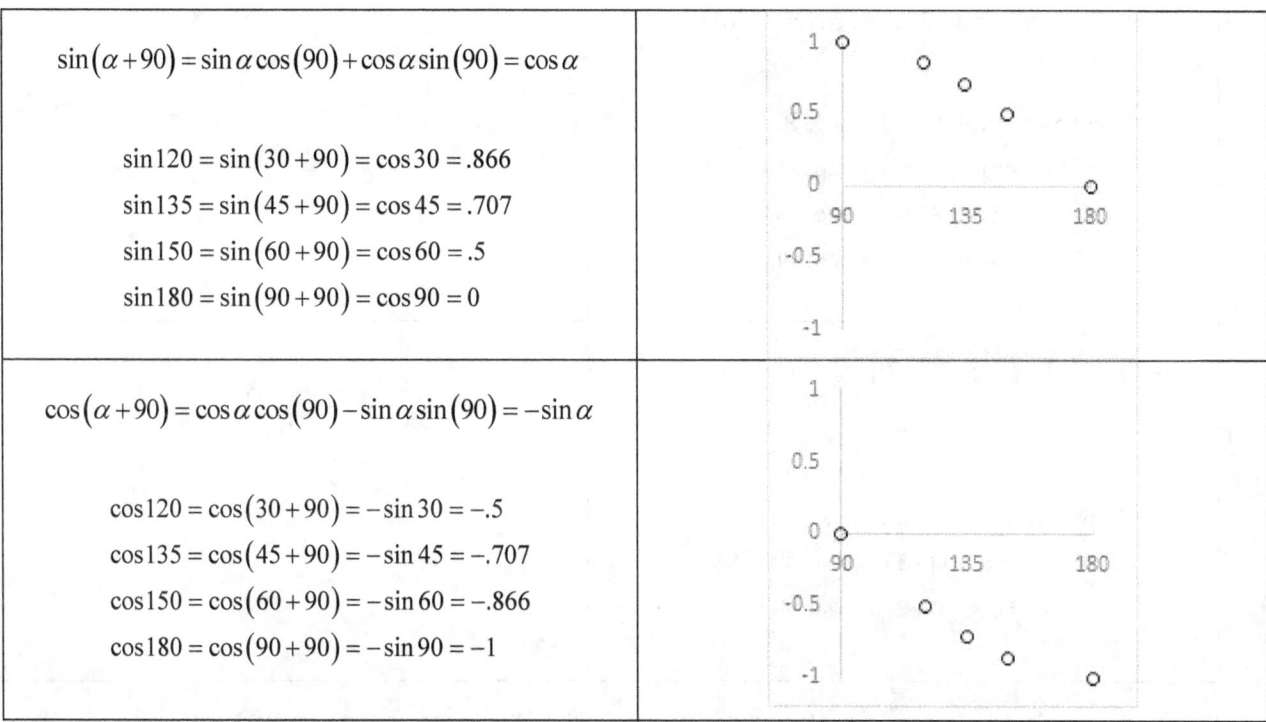

$$\sin(\alpha+90) = \sin\alpha\cos(90) + \cos\alpha\sin(90) = \cos\alpha$$

$$\sin 120 = \sin(30+90) = \cos 30 = .866$$
$$\sin 135 = \sin(45+90) = \cos 45 = .707$$
$$\sin 150 = \sin(60+90) = \cos 60 = .5$$
$$\sin 180 = \sin(90+90) = \cos 90 = 0$$

$$\cos(\alpha+90) = \cos\alpha\cos(90) - \sin\alpha\sin(90) = -\sin\alpha$$

$$\cos 120 = \cos(30+90) = -\sin 30 = -.5$$
$$\cos 135 = \cos(45+90) = -\sin 45 = -.707$$
$$\cos 150 = \cos(60+90) = -\sin 60 = -.866$$
$$\cos 180 = \cos(90+90) = -\sin 90 = -1$$

The following table shows the calculations that extend the sine and cosine to angles from 180 to 270 degrees. This is called Quadrant III.

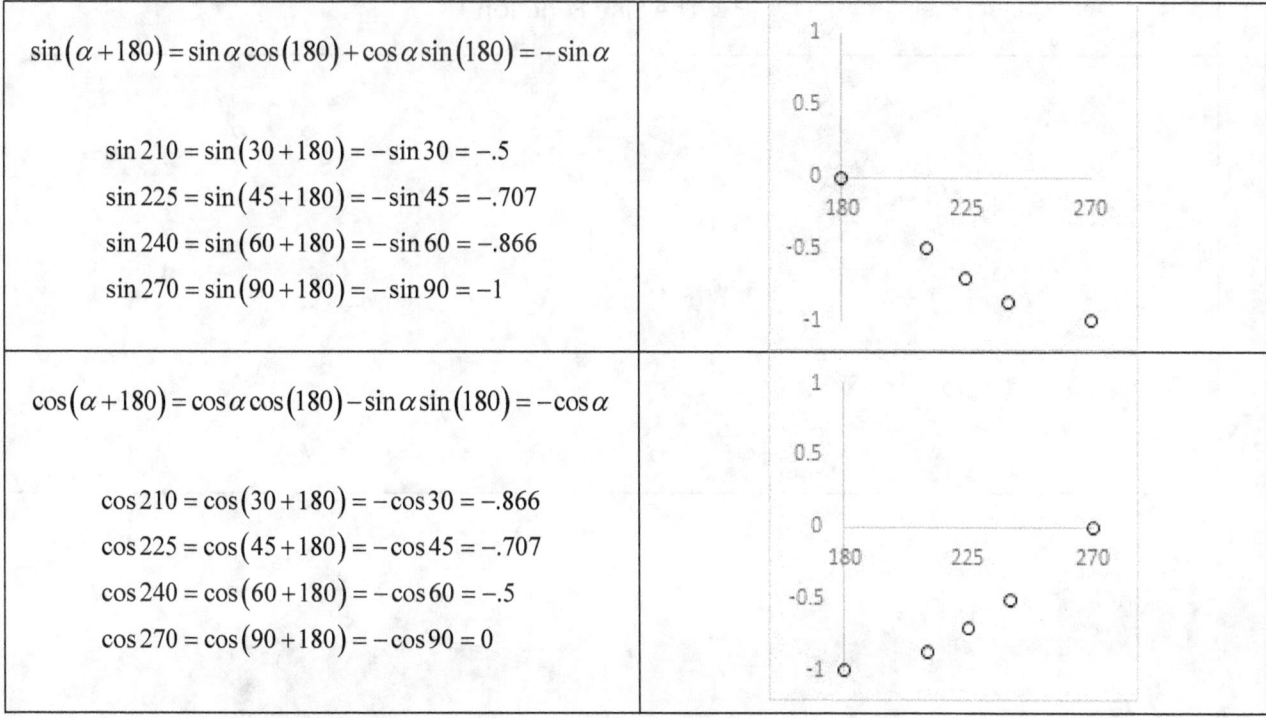

$$\sin(\alpha+180) = \sin\alpha\cos(180) + \cos\alpha\sin(180) = -\sin\alpha$$

$$\sin 210 = \sin(30+180) = -\sin 30 = -.5$$
$$\sin 225 = \sin(45+180) = -\sin 45 = -.707$$
$$\sin 240 = \sin(60+180) = -\sin 60 = -.866$$
$$\sin 270 = \sin(90+180) = -\sin 90 = -1$$

$$\cos(\alpha+180) = \cos\alpha\cos(180) - \sin\alpha\sin(180) = -\cos\alpha$$

$$\cos 210 = \cos(30+180) = -\cos 30 = -.866$$
$$\cos 225 = \cos(45+180) = -\cos 45 = -.707$$
$$\cos 240 = \cos(60+180) = -\cos 60 = -.5$$
$$\cos 270 = \cos(90+180) = -\cos 90 = 0$$

The following table shows the calculations that extend the sine and cosine to angles from 270 to 360 degrees. This is called Quadrant IV.

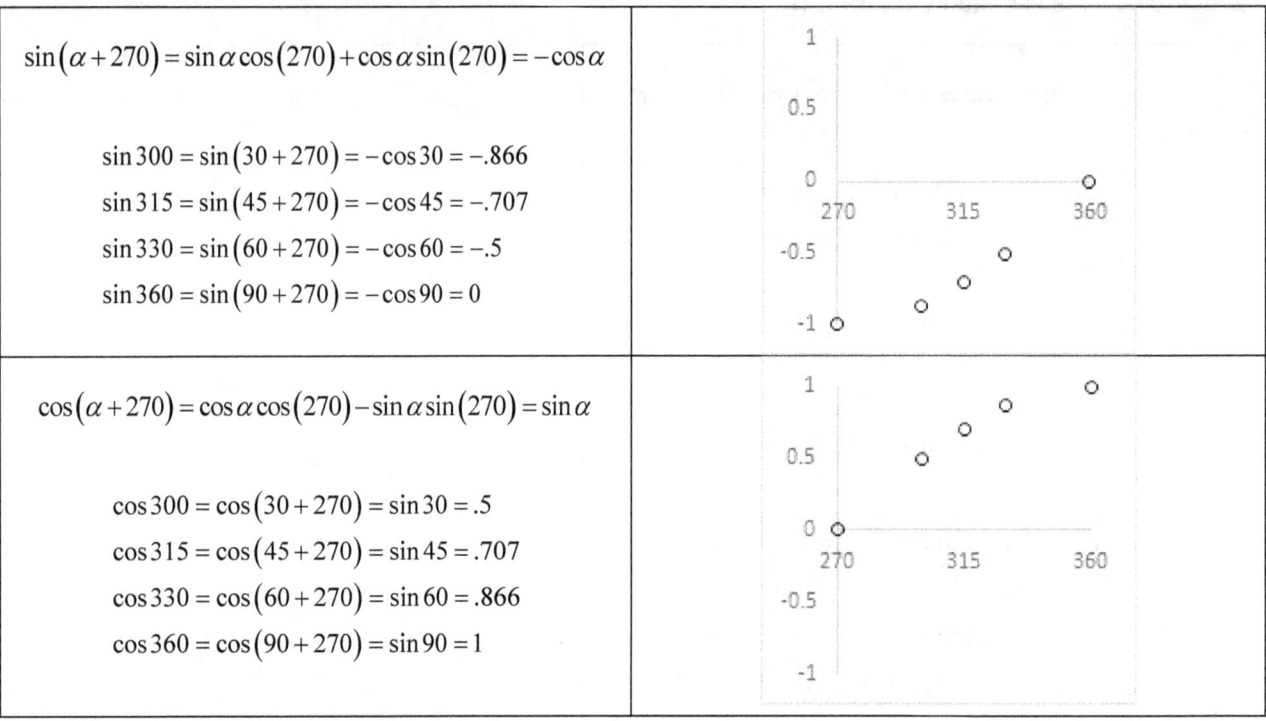

| $\sin(\alpha+270) = \sin\alpha\cos(270) + \cos\alpha\sin(270) = -\cos\alpha$ | |
| --- | --- |
| $\sin 300 = \sin(30+270) = -\cos 30 = -.866$ | |
| $\sin 315 = \sin(45+270) = -\cos 45 = -.707$ | |
| $\sin 330 = \sin(60+270) = -\cos 60 = -.5$ | |
| $\sin 360 = \sin(90+270) = -\cos 90 = 0$ | |
| $\cos(\alpha+270) = \cos\alpha\cos(270) - \sin\alpha\sin(270) = \sin\alpha$ | |
| $\cos 300 = \cos(30+270) = \sin 30 = .5$ | |
| $\cos 315 = \cos(45+270) = \sin 45 = .707$ | |
| $\cos 330 = \cos(60+270) = \sin 60 = .866$ | |
| $\cos 360 = \cos(90+270) = \sin 90 = 1$ | |

The following table is a composite of the sine function from Quadrants I to IV. Since it is cyclic, the composite is extended to the interval, −180 to 540 degrees.

| The Sine Function |
| --- |
| 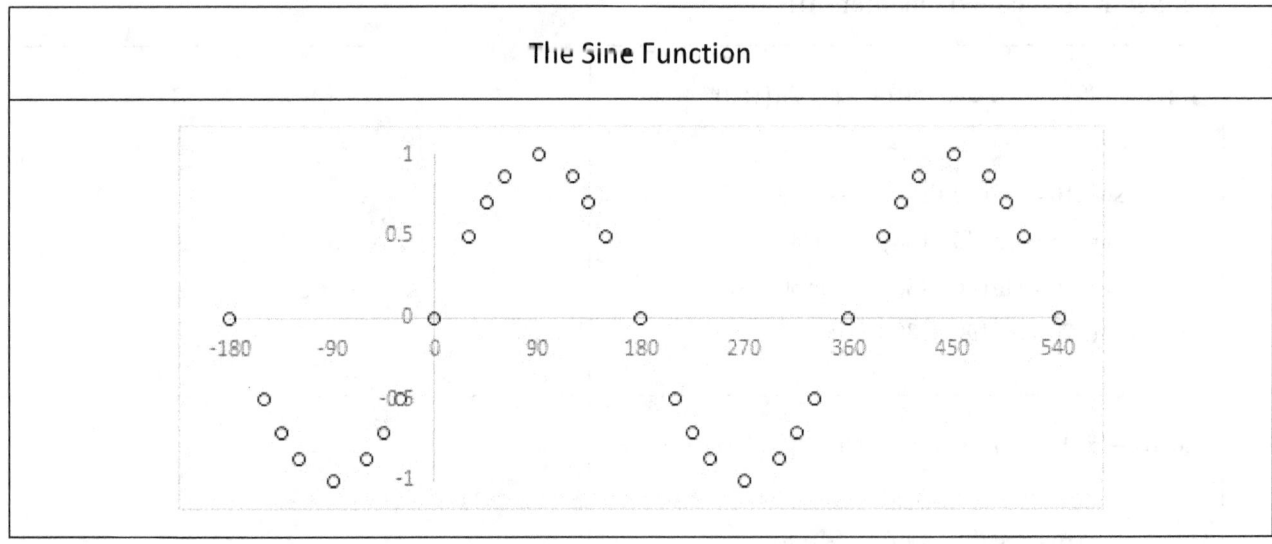 |

The following table is a composite of the cosine function from Quadrants I to IV. Since it is cyclic, the composite is extended to the interval, −180 to 540 degrees.

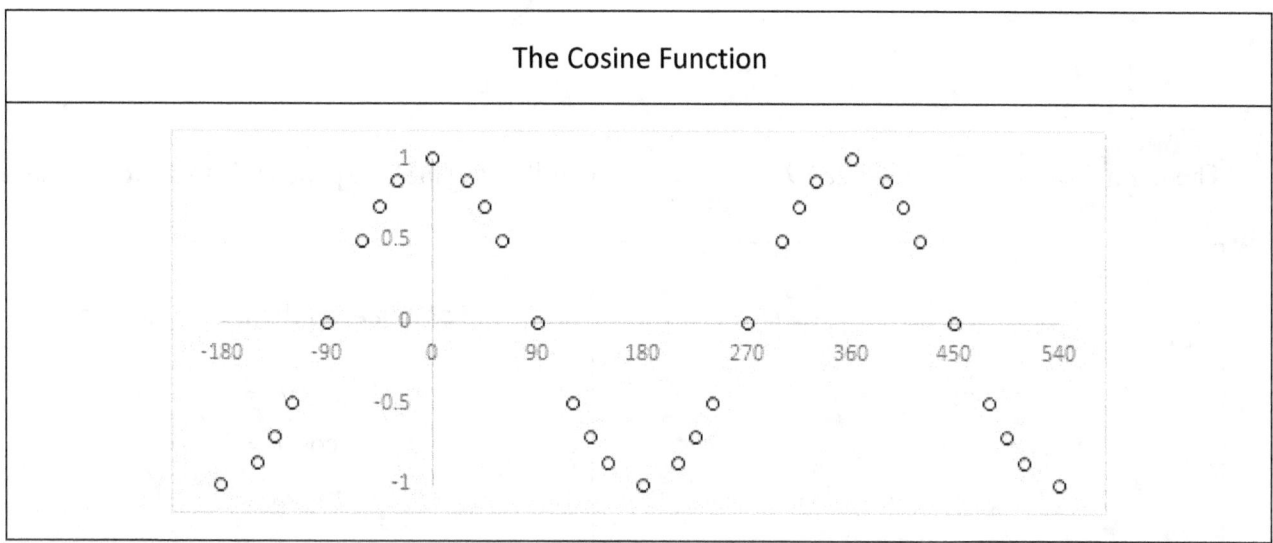

The tangent function is the ratio $\dfrac{\sin\theta}{\cos\theta}$. It can be constructed from the above tables. The dashed line shows what happens as $\cos\theta$ approaches zero.

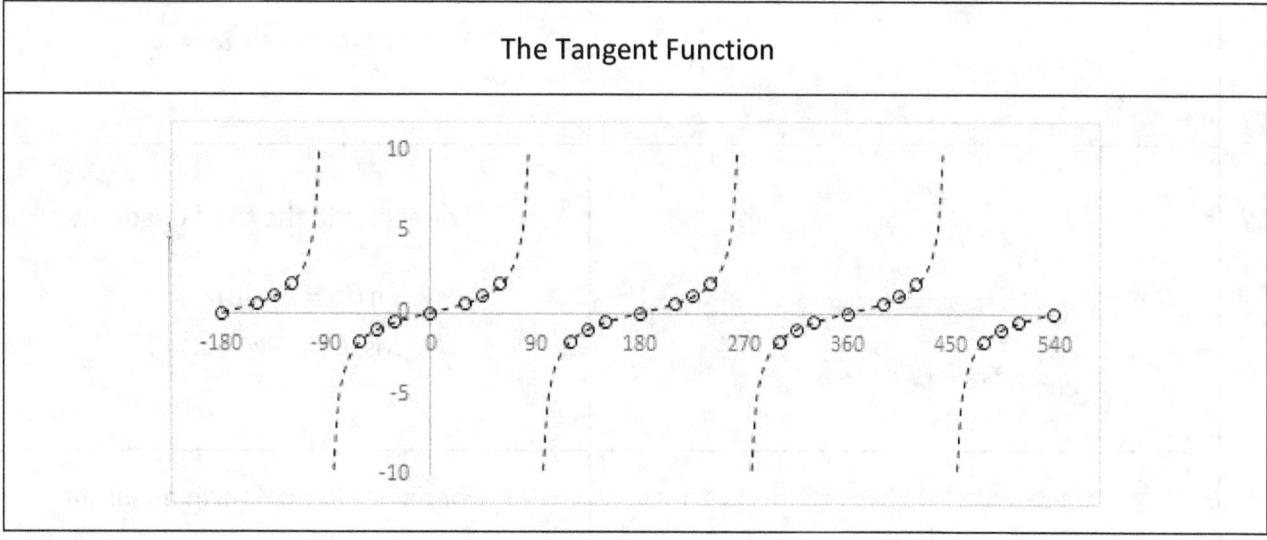

So far, we have defined the sine, cosine and tangent as being dimensionless ratios. Suppose we have a point P in quadrant I. Let's normalize it so that its distance to the origin is unity, that is, $r = 1$.

$$P(x,y) = \frac{P}{\sqrt{x^2 + y^2}}$$

Then, $\sin\theta = \dfrac{y}{r} = y$ and $\cos\theta = \dfrac{x}{r} = x$. The following table expands this to all four quadrants.

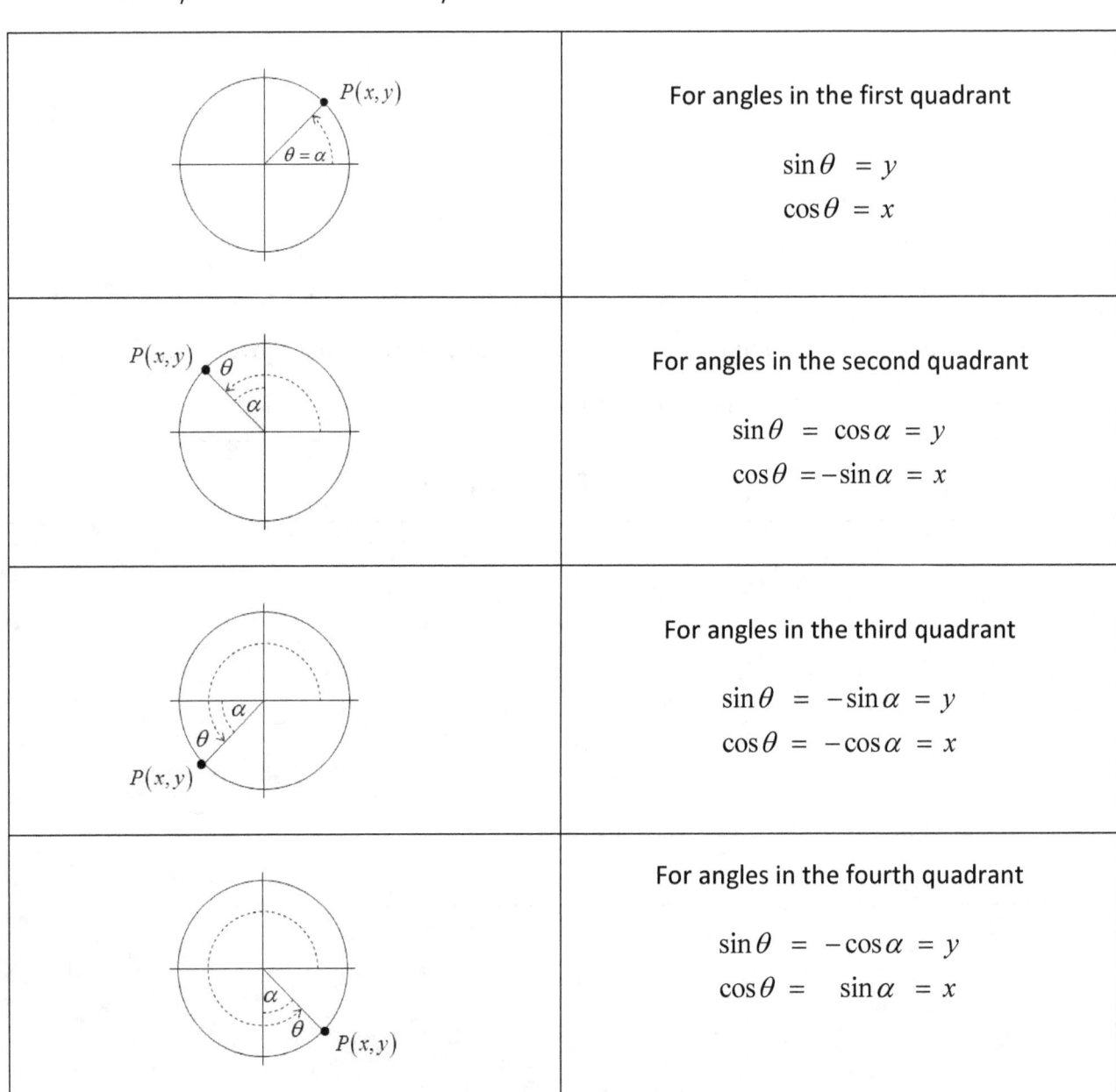

| | |
|---|---|
| | For angles in the first quadrant $$\sin\theta = y$$ $$\cos\theta = x$$ |
| | For angles in the second quadrant $$\sin\theta = \cos\alpha = y$$ $$\cos\theta = -\sin\alpha = x$$ |
| | For angles in the third quadrant $$\sin\theta = -\sin\alpha = y$$ $$\cos\theta = -\cos\alpha = x$$ |
| | For angles in the fourth quadrant $$\sin\theta = -\cos\alpha = y$$ $$\cos\theta = \sin\alpha = x$$ |

The conclusions are that the sine of any angle is its normalized $y$ component and the cosine is its normalized $x$ component. This automatically takes care of the sign.

## Computing the Sine at All Angles

Like $2^x$ and $\log_2(x)$ in chapter 2, the sine is a function that can be expanded in a Taylor series. The following begins a discussion for using this series to compute the sine at all angles.

The sine of an angle is a dimensionless quantity. Mathematicians use a dimensionless angle to compute this dimensionless quantity. They define the *radian* as the dimensionless angle that exists when the radius is equal to the arc that subtends the angle. See the following table.

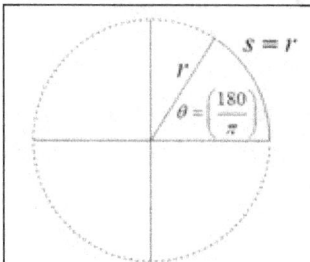

• From the equation of the circumference $2\pi r \rightarrow$ a 360 degree angle

• One *radian* is defined as the angle when $r = s$.

• Therefore, $2\pi * radian = 360^o$, or $radian = \dfrac{180^o}{\pi} = 57.296 \bullet\bullet\bullet$ degrees

With x as the angle in radians, the Taylor series for the sine function is

$$\text{sine}(x) = \sum_{n=1}^{\infty} (-1)^{n-1} \frac{x^{2n-1}}{(2n-1)!}$$

• The theory calls for an infinite series. Accuracy determines how many terms to use.
• The series is valid for all x. Therefore, the sine can be computed for all angles.
• The value of the constant term is $(2n-1)! = (2n-1)*(2n-2)* \bullet\bullet\bullet *3*2*1$.

This series will be compared to the very accurate VBA sine function. The syntax for this is sin(x). This VBA function may or may not use the Taylor series. But it is also just an approximation. The following table shows the comparison. The Taylor series (shown with circles) uses ten terms.

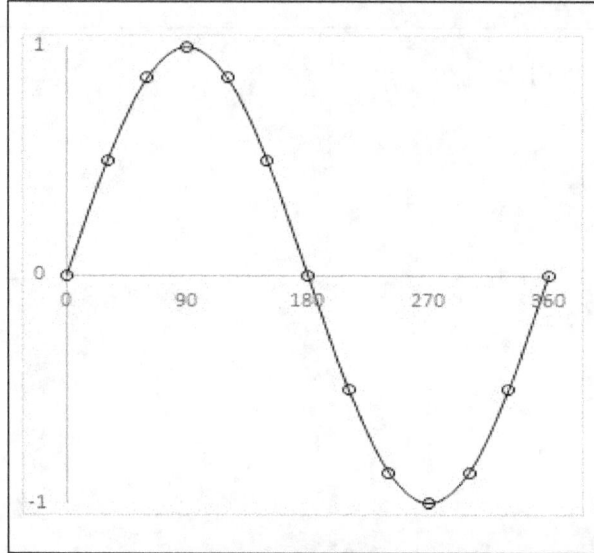

• The Taylor series that was used for the sine

$$\sum_{n=1}^{10} (-1)^{n-1} \frac{x^{2n-1}}{(2n-1)!}$$

• The first four terms

$$x - \frac{x^3}{3!} + \frac{x^5}{5!} - \frac{x^7}{7!}$$

• sin(-x) = -sin(x) for all angles

• sin(x) = x within about 1 percent for angles less than ten degrees

The simplified Taylor series compares remarkably well.

## Computing the Cosine at All Angles

The Taylor series for the cosine function is

$$\text{cosine}(\mathbf{x}) = \sum_{n=0}^{\infty}(-1)^n \frac{\mathbf{x}^{2n}}{(2n)!}$$

- Again, accuracy determines how many terms to use.
- The series is valid for all x. Therefore, the cosine can be computed for all angles.
- The value of the constant term is $(2n)! = (2n)*(2n-1)* \bullet \bullet \bullet *3*2*1$.

This series will be compared to the very accurate, though approximate, VBA cosine function. The syntax for this is **cos(x)**. This **VBA** function may or may not use the Taylor series. The following table shows the comparison. The Taylor series (shown with circles) uses eleven terms.

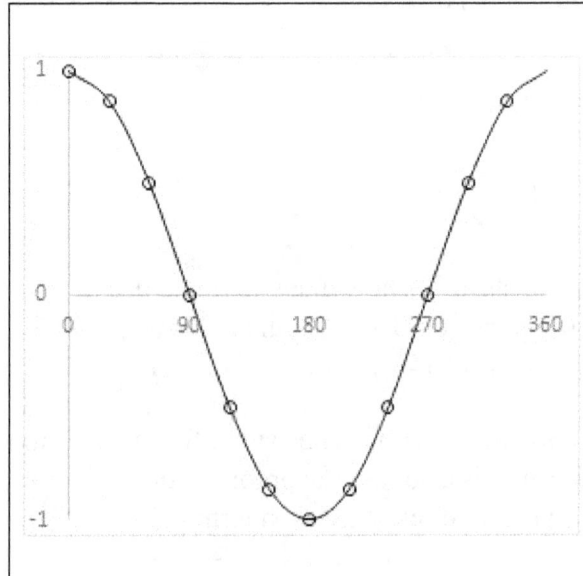

- The Taylor series that was used for the cosine

$$\sum_{n=0}^{10}(-1)^n \frac{\mathbf{x}^{2n}}{(2n)!}$$

- The first four terms

$$1 - \frac{\mathbf{x}^2}{2!} + \frac{\mathbf{x}^4}{4!} - \frac{\mathbf{x}^6}{6!}$$

- cos(-x) = cos(x) for all angles

- cos(x) = 1 within about 1 percent for angles less than eight degrees

The simplified Taylor series compares remarkably well.

## Computing the Tangent at All Angles

The Taylor series for the tangent can be computed from the series for the sine and cosine.

$$tangent(x) = \frac{sine(x)}{cosine(x)}$$

This series is valid for all x. Therefore, the tangent can be computed for all angles.

This series will be compared to the very accurate, though approximate, VBA tangent function. The syntax for this is tan(x). This VBA function may or may not use the Taylor series. The following table shows the comparison. The Taylor series is shown with bullets.

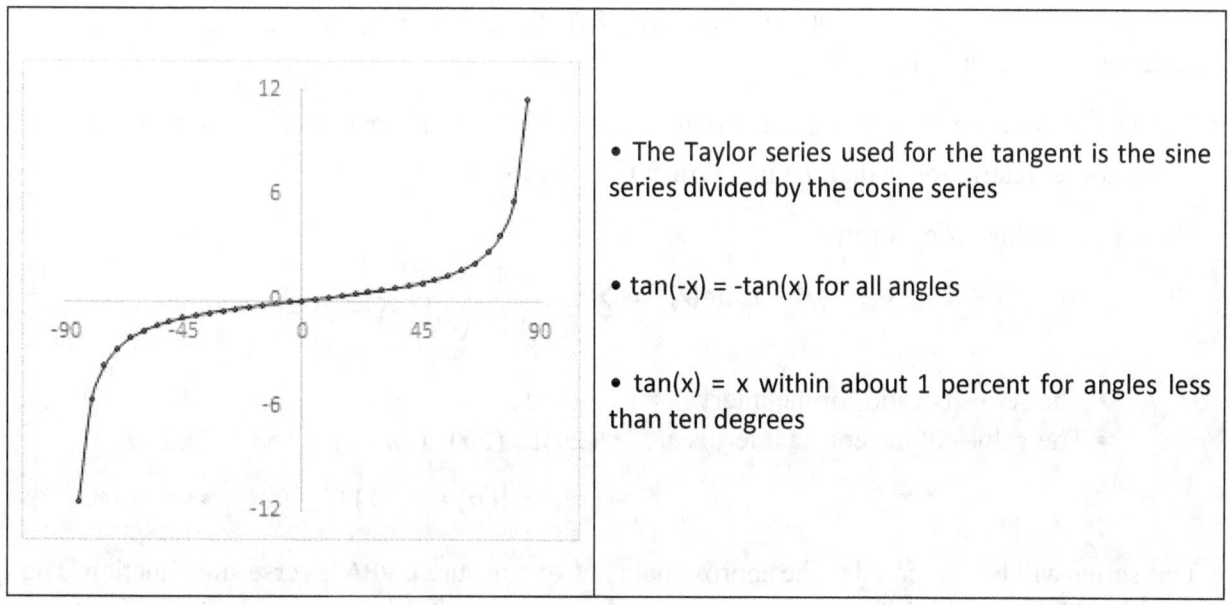

- The Taylor series used for the tangent is the sine series divided by the cosine series

- tan(-x) = -tan(x) for all angles

- tan(x) = x within about 1 percent for angles less than ten degrees

The simplified Taylor series compares remarkably well.

# Computing the Inverse Sine Function

Let's begin with a plot of the sine function.

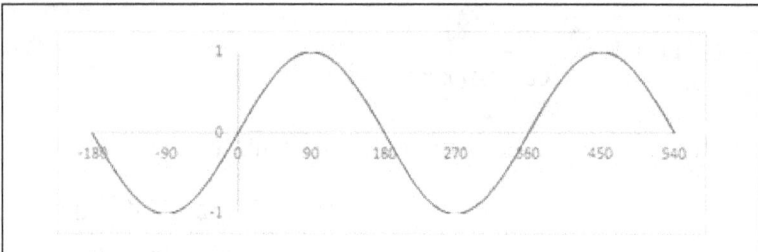

Define $\sin\theta = x$

where x is some number between $\pm 1$

The statement of the inverse problem is *given x, find $\theta$*. Stated mathematically, $\theta = \sin^{-1}(x)$. For example, if $x = 1$, $\theta$ can be 90, 450, and others. To be able to talk about a single inverse of the sine, there must be a restriction.

- The obvious restriction is that the output of $\sin^{-1}(x)$ will always be an angle between $\pm 90°$.
- Another restriction is that the input must be a number between $\pm 1$.

There is a Taylor series for this.

$$\text{asine(x)} = \sum_{n=0}^{\infty} \frac{(2n)!}{2^{2n}(n!)^2} \frac{x^{2n+1}}{(2n+1)}$$

- The series is valid for the interval $-1 < x < 1$.
- The values of the constant terms are: • $(2n)! = (2n)*(2n-1)* \bullet\bullet\bullet *3*2*1$;

  • $(n!)^2 - \{(n)*(n-1)*(n-2)* \bullet\bullet\bullet *3*2*1\}^2$.

This series will be compared to the approximate but very accurate **VBA** inverse sine function. The syntax for this is **Asin(x)**. This VBA function may or may not use the Taylor series. The following table shows the comparison. The Taylor series (shown with circles) uses eleven terms.

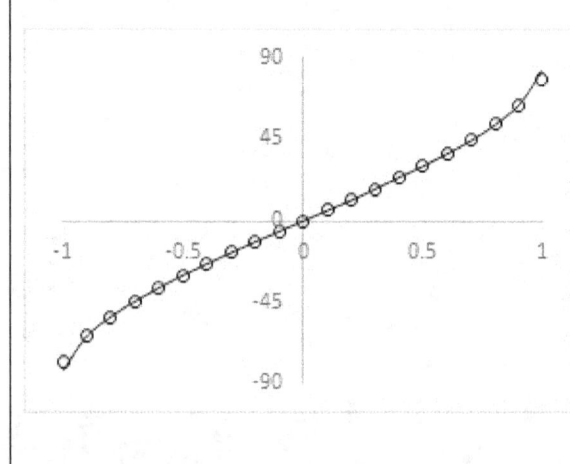

- The Taylor series used for the inverse sine

$$\sum_{n=0}^{10} \frac{(2n)!}{2^{2n}(n!)^2} \frac{x^{2n+1}}{(2n+1)}$$

- The first four terms

$$x + \frac{1}{2}*\frac{x^3}{3} + \frac{3}{8}*\frac{x^5}{5} + \frac{15}{48}*\frac{x^7}{7}$$

- $\sin^{-1}(-x) = -\sin^{-1}(x)$ for all angles in its range
- $\sin^{-1}(x) = x$ within about 1 percent for angles less than ten degrees

The simplified Taylor series compares remarkably well.

# Computing the Inverse Cosine Function

The following figure shows a plot of the cosine function.

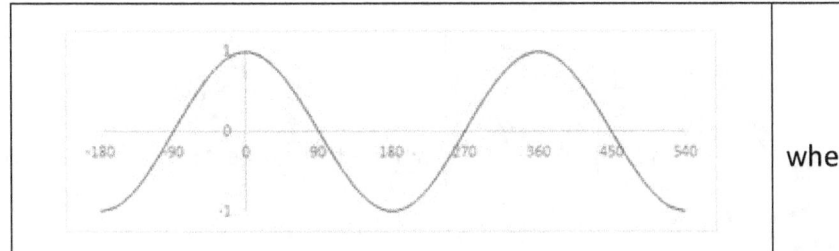

Define $\cos\theta = x$

where x is some number between $\pm1$

The statement of the inverse problem is *given x, find $\theta$*. Stated mathematically, $\theta = \cos^{-1}(x)$. For example, if $x = 1$, $\theta$ can be 0, 360, and others. To be able to talk about a single inverse of the cosine, there must be a restriction.

• The obvious restriction is that the output of $\cos^{-1}(x)$ will always be an angle between 0 and $180^0$.
• Another restriction is that the input must be a number between $\pm1$.

There is a Taylor series for this.

$$\text{acosine(x)} = \sum_{n=0}^{\infty}\left(\frac{\pi}{2} - \frac{(2n)!}{2^{2n}(n!)^2}\frac{x^{2n+1}}{(2n+1)}\right)$$

• The series is valid for the interval $-1 < x < 1$.
• The values of the constant terms are the same as those for the inverse sine function.

This series will be compared to the very accurate, though approximate, VBA inverse cosine function. The syntax for this is: Acos(x). This VBA function may or may not use the Taylor series. The following table shows the comparison. The Taylor series (shown with circles) uses eleven terms.

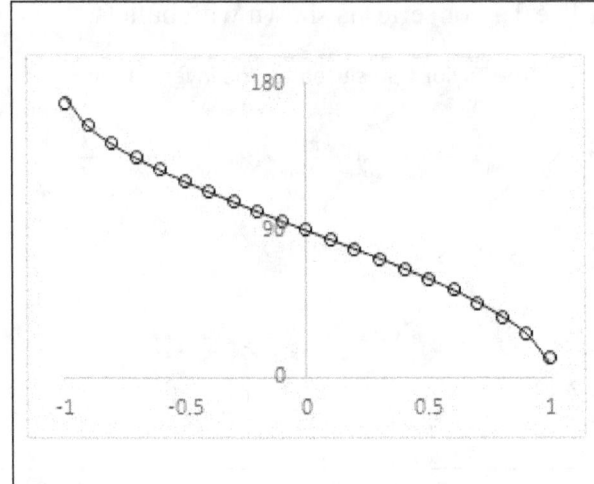

• The Taylor series used for the inverse cosine

$$\sum_{n=0}^{10}\left(\frac{\pi}{2} - \frac{(2n)!}{2^{2n}(n!)^2}\frac{x^{2n+1}}{(2n+1)}\right)$$

• The first five terms

$$\frac{\pi}{2} - x - \frac{1}{2}*\frac{x^3}{3} - \frac{3}{8}*\frac{x^5}{5} - \frac{15}{48}*\frac{x^7}{7}$$

The simplified Taylor series compares remarkably well.

## Computing the Inverse Tangent Function

The following figure shows a plot of the tangent.

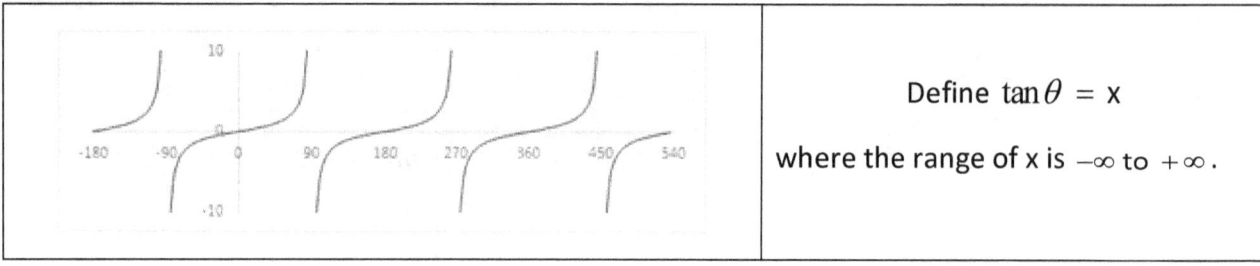

Define $\tan\theta = x$

where the range of $x$ is $-\infty$ to $+\infty$.

The statement of the inverse problem is *given x, find θ.* Stated mathematically, $\theta = \tan^{-1}(x)$. For example, if $x = 0$, $\theta$ can be $-180, 0, 180, 360$, and others. To be able to talk about a single inverse of the tangent, there must be a restriction. The restriction is that the output of $\tan^{-1}(x)$ will always be an angle between $\pm 90$.

There is a Taylor series for this. It is divided into the following areas.

$$\text{If } x < -1, \text{ then } \text{atangent}(x) = -\frac{\pi}{2} - \sum_{n=0}^{\infty}(-1)^n \frac{1}{(2n+1)x^{(2n+1)}}.$$

$$\text{If } (-1 < x < 1), \text{ then } \text{atangent}(x) = \sum_{n=0}^{\infty}(-1)^n \frac{x^{(2n+1)}}{(2n+1)}.$$

$$\text{If } x > 1, \text{ then } \text{atangent}(x) = \frac{\pi}{2} - \sum_{n=0}^{\infty}(-1)^n \frac{1}{(2n+1)x^{(2n+1)}}.$$

$$\text{If } x = 1, \text{ then } \text{atangent}(x) = \frac{\pi}{4} \quad \text{ or, if } x = -1, \text{ then } \text{atangent}(x) = -\frac{\pi}{4}.$$

This series will be compared to the very accurate VBA inverse tangent function. The syntax for this is Atn(x). The following table shows the comparison. The Taylor series is shown with bullets.

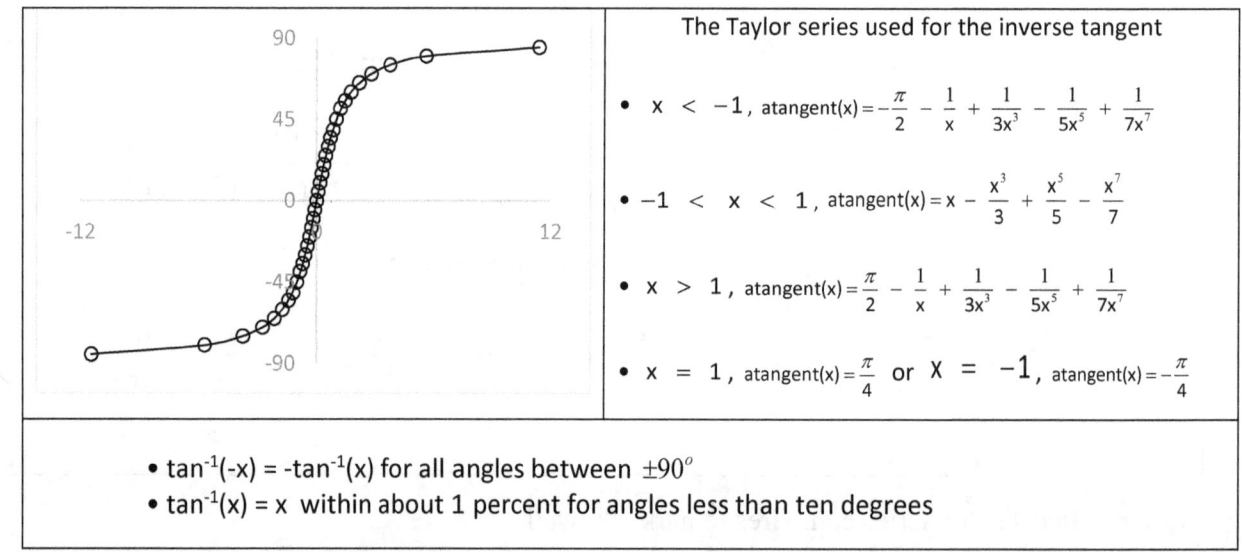

The Taylor series used for the inverse tangent

- $x < -1$, $\text{atangent}(x) = -\frac{\pi}{2} - \frac{1}{x} + \frac{1}{3x^3} - \frac{1}{5x^5} + \frac{1}{7x^7}$

- $-1 < x < 1$, $\text{atangent}(x) = x - \frac{x^3}{3} + \frac{x^5}{5} - \frac{x^7}{7}$

- $x > 1$, $\text{atangent}(x) = \frac{\pi}{2} - \frac{1}{x} + \frac{1}{3x^3} - \frac{1}{5x^5} + \frac{1}{7x^7}$

- $x = 1$, $\text{atangent}(x) = \frac{\pi}{4}$ or $x = -1$, $\text{atangent}(x) = -\frac{\pi}{4}$

- $\tan^{-1}(-x) = -\tan^{-1}(x)$ for all angles between $\pm 90°$
- $\tan^{-1}(x) = x$ within about 1 percent for angles less than ten degrees

## Analytic Trigonometry

The algebra of trigonometry uses equations that are called identities. The most widely used identity is

$$\sin^2 \alpha + \cos^2 \alpha = 1 .$$

This is derived by applying the Pythagorean theorem to a right triangle that has a hypotenuse of *one*.

This section shows the derivations of other useful identities.

- $\sin(\alpha+\beta)$ and $\cos(\alpha+\beta)$
- $\sin(\alpha-\beta)$ and $\cos(\alpha-\beta)$
- $\tan(\alpha+\beta)$ and $\tan(\alpha-\beta)$
- $\sin(2\alpha)$, $\cos(2\alpha)$, and $\tan(2\alpha)$
- $\sin^2(\alpha)$, $\cos^2(\alpha)$, and $\tan^2(\alpha)$
- $\sin(\alpha/2)$, $\cos(\alpha/2)$, and $\tan(\alpha/2)$
- $\cos\alpha\cos\beta$, $\sin\alpha\sin\beta$, $\sin\alpha\cos\beta$, and $\cos\alpha\sin\beta$
- $\cos\alpha + \cos\beta$, $\cos\alpha - \cos\beta$, $\sin\alpha + \sin\beta$, and $\sin\alpha - \sin\beta$

Earlier in this chapter, we derived $\sin(\alpha+\beta)$ and $\cos(\alpha+\beta)$. The following table is a repeat of this derivation.

Deriving Formulas for $\sin(\alpha+\beta)$ and $\cos(\alpha+\beta)$.

Note: The length of line $\overline{OC}$ is unity.

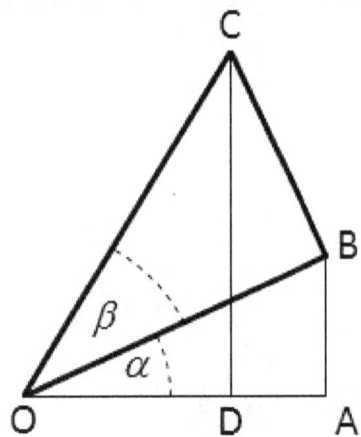

1. $\sin(\alpha+\beta) = \overline{CD}$

2. $\cos(\alpha+\beta) = \overline{OD} = \overline{OA} - \overline{AD}$

- $\sin\beta = \overline{BC}$
- $\cos\beta = \overline{OB}$
- $\sin\alpha = \dfrac{\overline{AB}}{\overline{OB}}$

  Hence $\rightarrow$ $\overline{AB} = \overline{OB}\sin\alpha = \sin\alpha\cos\beta$

- $\cos\alpha = \dfrac{\overline{OA}}{\overline{OB}}$

  Hence $\rightarrow$ $\overline{OA} = \overline{OB}\cos\alpha = \cos\alpha\cos\beta$

Adding Line $\overline{BE}$ So That $\overline{CD} = \overline{DE} + \overline{CE}$ and $\overline{DE} = \overline{AB}$ and $\overline{AD} = \overline{BE}$

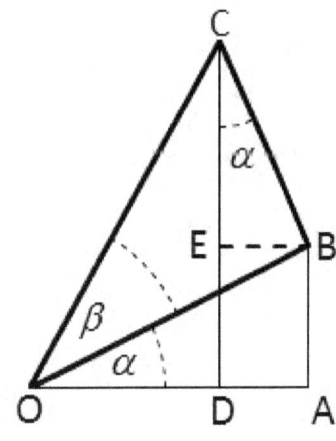

- $\cos\alpha = \dfrac{\overline{CE}}{\overline{BC}}$

  Hence $\rightarrow$ $\overline{CE} = \overline{BC}\cos\alpha = \cos\alpha\sin\beta$

- $\sin\alpha = \dfrac{\overline{BE}}{\overline{BC}}$

  Hence $\rightarrow$ $\overline{BE} = \overline{BC}\sin\alpha = \sin\alpha\sin\beta$

Combining the above equations

1. $\sin(\alpha+\beta) = \overline{CD} = \overline{DE} + \overline{CE} = \overline{AB} + \overline{CE} = \sin\alpha\cos\beta + \cos\alpha\sin\beta$ $\rightarrow$ $\boxed{\sin(\alpha+\beta) = \sin\alpha\cos\beta + \cos\alpha\sin\beta}$

2. $\cos(\alpha+\beta) = \overline{OD} = \overline{OA} - \overline{AD} = \overline{OA} - \overline{BE} = \cos\alpha\cos\beta - \sin\alpha\sin\beta$ $\rightarrow$ $\boxed{\cos(\alpha+\beta) = \cos\alpha\cos\beta - \sin\alpha\sin\beta}$

We have previously used $\sin(\alpha+\beta)$ and $\cos(\alpha+\beta)$ to compute the sine and cosine in all four quadrants. The following table summarizes this.

| Reducing All Angles to Acute Angles Using the Formulas $\sin(\alpha+\beta)$ and $\cos(\alpha+\beta)$ | |
|---|---|
| $\sin(\theta+90) = \sin\theta\cos(90) + \cos\theta\sin(90) = \cos\theta$ <br><br> $\boxed{\sin(\theta+90) = \cos\theta}$ | $\cos(\theta+90) = \cos\theta\cos(90) - \sin\theta\sin(90) = -\sin\theta$ <br><br> $\boxed{\cos(\theta+90) = -\sin\theta}$ |

- $\sin(\theta+180) = \sin\left(\underbrace{\theta+90}_{\alpha}+90\right) = \sin\alpha\cos(90) + \cos\alpha\sin(90) = \cos(\theta+90) = -\sin\theta$

$$\boxed{\sin(\theta+180) = -\sin\theta}$$

- $\cos(\theta+180) = \cos\left(\underbrace{\theta+90}_{\alpha}+90\right) = \cos\alpha\cos(90) - \sin\alpha\sin(90) = -\sin(\theta+90) = -\cos\theta$

$$\boxed{\cos(\theta+180) = -\cos\theta}$$

- $\sin(\theta+270) = \sin\left(\underbrace{\theta+180}_{\alpha}+90\right) = \sin\alpha\cos(90) + \cos\alpha\sin(90) = \cos(\theta+180) = -\cos\theta$

$$\boxed{\sin(\theta+270) = -\cos\theta}$$

- $\cos(\theta+270) = \cos\left(\underbrace{\theta+180}_{\alpha}+90\right) = \cos\alpha\cos(90) - \sin\alpha\sin(90) = -\sin(\theta+180) = \sin\theta$

$$\boxed{\cos(\theta+270) = \sin\theta}$$

In the following table, we derive sin(α–β) and cos(α–β) from sin(α+β) and cos(α+β).

| Deriving Formulas for $\sin(\alpha-\beta)$ and $\cos(\alpha-\beta)$ |
|---|
| Starting from $\sin(\alpha+\beta)=\sin\alpha\cos\beta+\cos\alpha\sin\beta$, substitute $(-\beta)$ for $\beta$. <br> • $\sin(\alpha+(-\beta))=\sin\alpha\cos(-\beta)+\cos\alpha\sin(-\beta)$ → $\boxed{\sin(\alpha-\beta)=\sin\alpha\cos\beta-\cos\alpha\sin\beta}$ <br><br> Starting from $\cos(\alpha+\beta)=\cos\alpha\cos\beta-\sin\alpha\sin\beta$, substitute $(-\beta)$ for $\beta$. <br> • $\cos(\alpha+(-\beta))=\cos\alpha\cos(-\beta)-\sin\alpha\sin(-\beta)$ → $\boxed{\cos(\alpha-\beta)=\cos\alpha\cos\beta+\sin\alpha\sin\beta}$ |

The following table uses sin(α–β) and cos(α–β) to compute the sine and cosine in all four quadrants.

| Reducing All Angles To Acute Angles Using the Formulas sin(α–β) and cos(α–β) | |
|---|---|
| $\sin(\theta-90)=\sin\theta\cos(90)-\cos\theta\sin(90)=-\cos\theta$ <br> $\boxed{\sin(\theta-90)=-\cos\theta}$ | $\cos(\theta-90)=\cos\theta\cos(90)+\sin\theta\sin(90)=\sin\theta$ <br> $\boxed{\cos(\theta-90)=\sin\theta}$ |
| $\boxed{\sin(\theta-180)=-\sin\theta}$  $\boxed{\cos(\theta-180)=-\cos\theta}$ | $\boxed{\sin(\theta-270)=\cos\theta}$  $\boxed{\cos(\theta-270)=-\sin\theta}$ |

In the following table, we show the derivation of tan(α+β) from sin(α+β) and cos(α+β). We also show the derivation of tan(α–β).

$$\tan(\alpha+\beta)=\frac{\sin(\alpha+\beta)}{\cos(\alpha+\beta)}=\frac{\sin\alpha\cos\beta+\cos\alpha\sin\beta}{\cos\alpha\cos\beta-\sin\alpha\sin\beta}=\frac{\frac{\sin\alpha\cos\beta+\cos\alpha\sin\beta}{\cos\alpha\cos\beta}}{\frac{\cos\alpha\cos\beta-\sin\alpha\sin\beta}{\cos\alpha\cos\beta}}=\frac{\frac{\sin\alpha}{\cos\alpha}+\frac{\sin\beta}{\cos\beta}}{1-\frac{\sin\alpha}{\cos\alpha}\frac{\sin\beta}{\cos\beta}}$$

$$\boxed{\tan(\alpha+\beta)=\frac{\tan\alpha+\tan\beta}{1-\tan\alpha\tan\beta}}$$

$$\tan(\alpha-\beta)=\frac{\tan\alpha+\tan(-\beta)}{1-\tan\alpha\tan(-\beta)}$$

$$\boxed{\tan(\alpha-\beta)=\frac{\tan\alpha-\tan\beta}{1+\tan\alpha\tan\beta}}$$

The following table derives the double-angle formulas, that is, $\sin(2\alpha)$, $\cos(2\alpha)$, and $\tan(2\alpha)$.

| |
|---|
| Since $\sin(\alpha+\beta)=\sin\alpha\cos\beta+\cos\alpha\sin\beta$ , then $\sin(\alpha+\alpha)=\sin\alpha\cos\alpha+\cos\alpha\sin\alpha$ , or $$\boxed{\sin(2\alpha)=2\sin\alpha\cos\alpha}$$ |
| Since $\cos(\alpha+\beta)=\cos\alpha\cos\beta-\sin\alpha\sin\beta$, then $\cos(2\alpha)=\cos^2\alpha-\sin^2\alpha=\cos^2\alpha-\left(1-\cos^2\alpha\right)$ . Therefore, $\boxed{\cos(2\alpha)=2\cos^2\alpha-1}$ or $\boxed{\cos(2\alpha)=1-2\sin^2\alpha}$ |
| Since $\tan(\alpha+\beta)=\dfrac{\tan\alpha+\tan\beta}{1-\tan\alpha\tan\beta}$ , then $\boxed{\tan(2\alpha)=\dfrac{2\tan\alpha}{1-\tan^2\alpha}}$ |

In chapter 7 on the conics, we encounter the following problem. Find $\alpha$ when we are given: $2\sin\alpha\cos\alpha=K\left(\cos^2\alpha-\sin^2\alpha\right)$ and K is a constant. From the above table

$$2\sin\alpha\cos\alpha=\sin(2\alpha) \text{ and } \cos(2\alpha)=\left(\cos^2\alpha-\sin^2\alpha\right).$$

Therefore, $\sin(2\alpha)=K\cos(2\alpha)$ or $\tan(2\alpha)=\dfrac{\sin(2\alpha)}{\cos(2\alpha)}=K$ . Finally, $\alpha=0.5\tan^{-1}(K)$ .

The following table shows the derivation of $\sin^2\alpha$ , $\cos^2\alpha$ , and $\tan^2\alpha$ .

| | |
|---|---|
| Since $\cos(2\alpha)=2\cos^2\alpha-1$, then $$\boxed{\cos^2\alpha=\dfrac{1+\cos(2\alpha)}{2}}$$ | Since $\cos(2\alpha)=1-2\sin^2\alpha$ , then $$\boxed{\sin^2\alpha=\dfrac{1-\cos(2\alpha)}{2}}$$ |
| Since $\tan^2\alpha=\dfrac{\sin^2\alpha}{\cos^2\alpha}=\dfrac{\dfrac{1-\cos(2\alpha)}{2}}{\dfrac{1+\cos(2\alpha)}{2}}$ , then $\boxed{\tan^2\alpha=\dfrac{1-\cos(2\alpha)}{1+\cos(2\alpha)}}$ . | |

These are frequently used with $\sin^2\alpha+\cos^2\alpha=1$ .

The following table shows the derivation of sin(α/2), cos(α/2), and tan(α/2).

| | | |
|---|---|---|
| Since $\cos^2 \alpha = \dfrac{1+\cos(2\alpha)}{2}$, | Since $\sin^2 \alpha = \dfrac{1-\cos(2\alpha)}{2}$, | Since $\tan^2 \alpha = \dfrac{1-\cos(2\alpha)}{1+\cos(2\alpha)}$, |
| then $\cos^2 \dfrac{\alpha}{2} = \dfrac{1+\cos\left(2\dfrac{\alpha}{2}\right)}{2}$. | then $\sin^2 \dfrac{\alpha}{2} = \dfrac{1-\cos\left(2\dfrac{\alpha}{2}\right)}{2}$. | then $\tan^2 \dfrac{\alpha}{2} = \dfrac{1-\cos(\alpha)}{1+\cos(\alpha)}$. |
| $\boxed{\cos\left(\dfrac{\alpha}{2}\right) = \pm\sqrt{\dfrac{1+\cos\alpha}{2}}}$ | $\boxed{\sin\dfrac{\alpha}{2} = \pm\sqrt{\dfrac{1-\cos\alpha}{2}}}$ | $\boxed{\tan\dfrac{\alpha}{2} = \pm\sqrt{\dfrac{1-\cos\alpha}{1+\cos\alpha}}}$ |

The Formula for $\tan\dfrac{\alpha}{2}$ Has Alternate Forms

| | |
|---|---|
| Conjugating the denominator of $\tan\dfrac{\alpha}{2} = \pm\sqrt{\dfrac{1-\cos\alpha}{1+\cos\alpha}}$, | Conjugating the numerator of $\tan\dfrac{\alpha}{2} = \pm\sqrt{\dfrac{1-\cos\alpha}{1+\cos\alpha}}$, |
| $\tan\dfrac{\alpha}{2} = \pm\sqrt{\dfrac{(1-\cos\alpha)(1-\cos\alpha)}{(1+\cos\alpha)(1-\cos\alpha)}} = \pm\sqrt{\dfrac{(1-\cos\alpha)^2}{(1-\cos^2\alpha)}} = \pm\sqrt{\dfrac{(1-\cos\alpha)^2}{\sin^2\alpha}}$ | $\tan\dfrac{\alpha}{2} = \pm\sqrt{\dfrac{(1-\cos\alpha)(1+\cos\alpha)}{(1+\cos\alpha)(1+\cos\alpha)}} = \pm\sqrt{\dfrac{(1-\cos^2\alpha)}{(1+\cos\alpha)^2}} = \pm\sqrt{\dfrac{\sin^2\alpha}{(1+\cos\alpha)^2}}$ |
| $\tan\dfrac{\alpha}{2} = \pm\dfrac{\lvert 1-\cos\alpha\rvert}{\lvert\sin\alpha\rvert}$ | $\tan\dfrac{\alpha}{2} = \pm\dfrac{\lvert\sin\alpha\rvert}{\lvert 1+\cos\alpha\rvert}$ |
| Because $\lvert 1-\cos\alpha\rvert \geq 0$, and $\tan\left(\dfrac{\alpha}{2}\right)$ and $\sin\alpha$ have the same sign, | Because $\lvert 1+\cos\alpha\rvert \geq 0$, and $\tan\left(\dfrac{u}{2}\right)$ and $\sin\alpha$ have the same sign, |
| $\boxed{\tan\dfrac{\alpha}{2} = \dfrac{1-\cos\alpha}{\sin\alpha}}$ | $\boxed{\tan\dfrac{\alpha}{2} = \dfrac{\sin\alpha}{1+\cos\alpha}}$ |

The following table shows the derivation of the so-called *product-to-sum* identities.

| Deriving the Product-to-Sum Identities | |
|---|---|
| We have previously derived<br>• $\cos(\alpha+\beta)=\cos\alpha\cos\beta-\sin\alpha\sin\beta$<br>• $\cos(\alpha-\beta)=\cos\alpha\cos\beta+\sin\alpha\sin\beta$ | We have previously derived<br>• $\sin(\alpha+\beta)=\sin\alpha\cos\beta+\cos\alpha\sin\beta$<br>• $\sin(\alpha-\beta)=\sin\alpha\cos\beta-\cos\alpha\sin\beta$ |
| If we add the above equations,<br>$$\cos(\alpha+\beta)+\cos(\alpha-\beta)=2\cos\alpha\cos\beta$$<br>$$\boxed{2\cos\alpha\cos\beta=\cos(\alpha+\beta)+\cos(\alpha-\beta)}$$ | If we add the above equations,<br>$$\sin(\alpha+\beta)+\sin(\alpha-\beta)=2\sin\alpha\cos\beta$$<br>$$\boxed{2\sin\alpha\cos\beta=\sin(\alpha+\beta)+\sin(\alpha-\beta)}$$ |
| If we subtract the above equations,<br>$$\cos(\alpha+\beta)-\cos(\alpha-\beta)=-2\sin\alpha\sin\beta$$<br>$$\boxed{2\sin\alpha\sin\beta=-\cos(\alpha+\beta)+\cos(\alpha-\beta)}$$ | If we subtract the above equations,<br>$$\sin(\alpha+\beta)-\sin(\alpha-\beta)=2\cos\alpha\sin\beta$$<br>$$\boxed{2\cos\alpha\sin\beta=\sin(\alpha+\beta)-\sin(\alpha-\beta)}$$ |

The following table shows the derivation of the so-called *sum-to-product* identities.

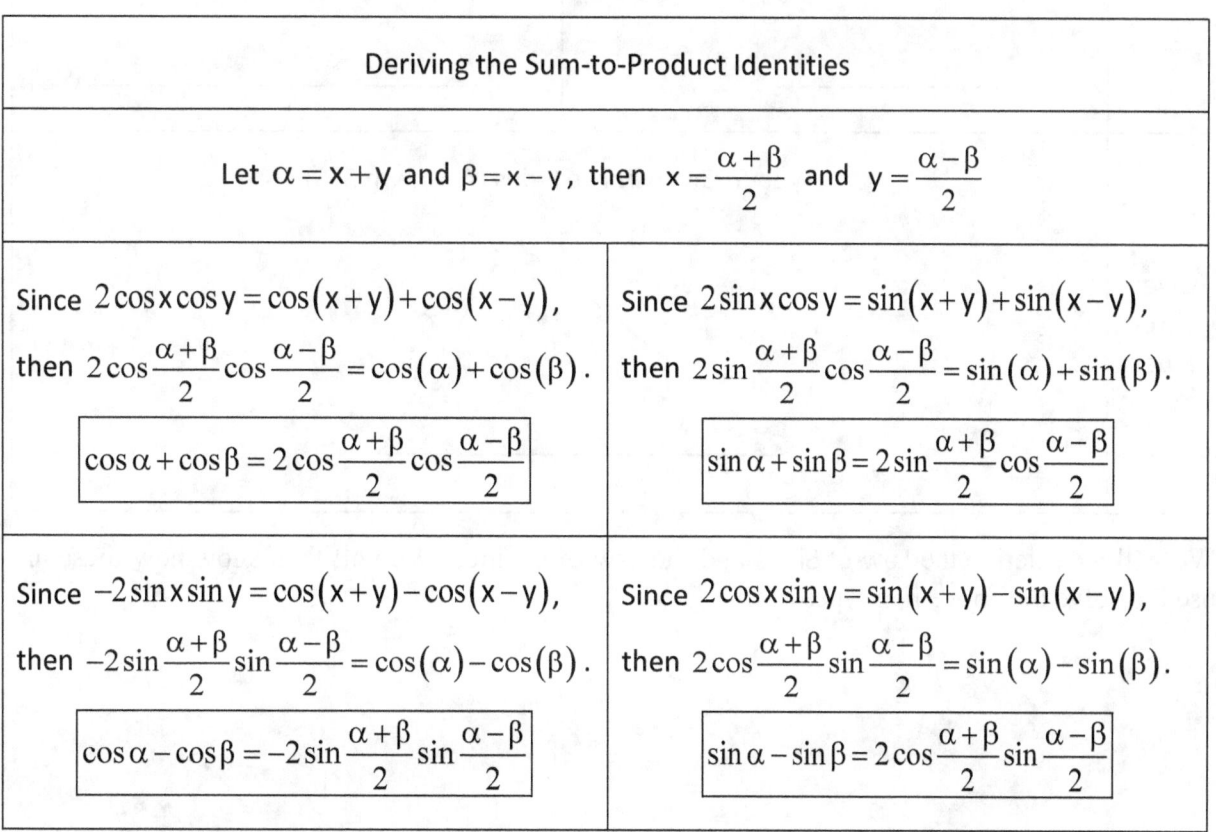

| Deriving the Sum-to-Product Identities | |
|---|---|
| Let $\alpha=x+y$ and $\beta=x-y$, then $x=\dfrac{\alpha+\beta}{2}$ and $y=\dfrac{\alpha-\beta}{2}$ | |
| Since $2\cos x\cos y=\cos(x+y)+\cos(x-y)$,<br>then $2\cos\dfrac{\alpha+\beta}{2}\cos\dfrac{\alpha-\beta}{2}=\cos(\alpha)+\cos(\beta)$.<br>$$\boxed{\cos\alpha+\cos\beta=2\cos\dfrac{\alpha+\beta}{2}\cos\dfrac{\alpha-\beta}{2}}$$ | Since $2\sin x\cos y=\sin(x+y)+\sin(x-y)$,<br>then $2\sin\dfrac{\alpha+\beta}{2}\cos\dfrac{\alpha-\beta}{2}=\sin(\alpha)+\sin(\beta)$.<br>$$\boxed{\sin\alpha+\sin\beta=2\sin\dfrac{\alpha+\beta}{2}\cos\dfrac{\alpha-\beta}{2}}$$ |
| Since $-2\sin x\sin y=\cos(x+y)-\cos(x-y)$,<br>then $-2\sin\dfrac{\alpha+\beta}{2}\sin\dfrac{\alpha-\beta}{2}=\cos(\alpha)-\cos(\beta)$.<br>$$\boxed{\cos\alpha-\cos\beta=-2\sin\dfrac{\alpha+\beta}{2}\sin\dfrac{\alpha-\beta}{2}}$$ | Since $2\cos x\sin y=\sin(x+y)-\sin(x-y)$,<br>then $2\cos\dfrac{\alpha+\beta}{2}\sin\dfrac{\alpha-\beta}{2}=\sin(\alpha)-\sin(\beta)$.<br>$$\boxed{\sin\alpha-\sin\beta=2\cos\dfrac{\alpha+\beta}{2}\sin\dfrac{\alpha-\beta}{2}}$$ |

## Solving a Triangle

Solving a triangle means to compute all of its angles and sides. A unique triangle cannot be specified by angles alone. You must know at least one side. The following table summarizes the minimum requirements that must be given to define a triangle.

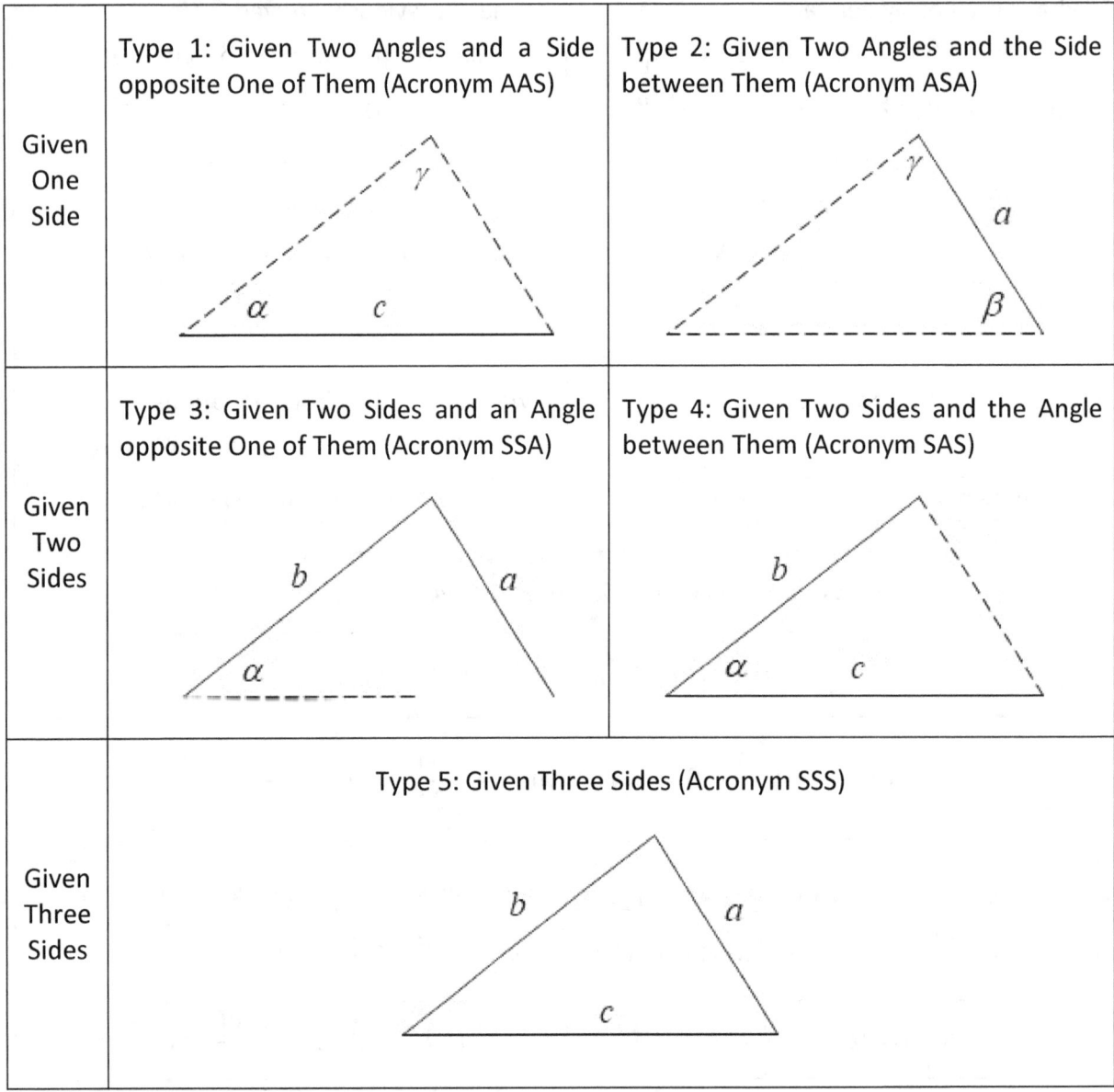

| | | |
|---|---|---|
| **Given One Side** | Type 1: Given Two Angles and a Side opposite One of Them (Acronym AAS) | Type 2: Given Two Angles and the Side between Them (Acronym ASA) |
| **Given Two Sides** | Type 3: Given Two Sides and an Angle opposite One of Them (Acronym SSA) | Type 4: Given Two Sides and the Angle between Them (Acronym SAS) |
| **Given Three Sides** | Type 5: Given Three Sides (Acronym SSS) | |

We will now derive the Law of Sines and the Law of Cosines. We will then show how these are used to solve the above five types of triangles.

The **Law of Sines** is derived in the following table.

| Deriving the Law of Sines Based on the Area of a Triangle $\left(\dfrac{1}{2}\text{base}*\text{height}\right)$ | |
|---|---|
| 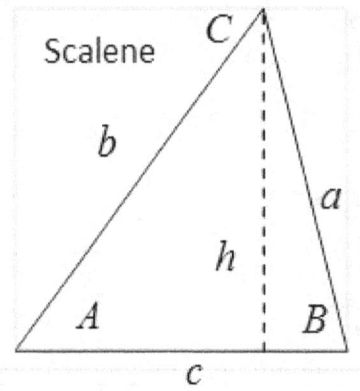 | • Angles are A, B and C. Sides are $a$, $b$ and $c$.<br><br>1. If height $h = a * \sin B$ and base is $c$, then $$Area = \frac{1}{2}ca\sin B$$<br>2. If height $h = b * \sin A$ and base is $c$, then $$Area = \frac{1}{2}cb\sin A$$<br>3. If height is perpendicular to base $b$ where $h = a * \sin C$, then $$Area = \frac{1}{2}ba\sin C$$ |
| 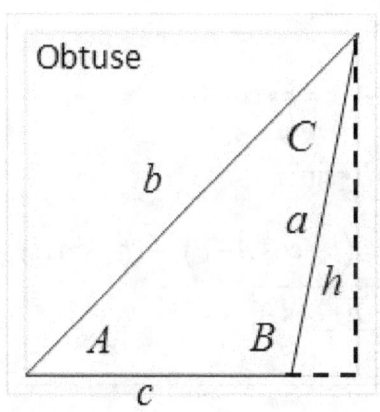 | 1. If $c$ is the base, then $h = a * \sin(180 - B) = a * \sin B$ $$Area = \frac{1}{2}ca\sin B$$<br>2. If height $h = b * \sin A$ and base is $c$, then $$Area = \frac{1}{2}cb\sin A$$<br>3. If height is perpendicular to base $b$ where $h = a * \sin C$, then $$Area = \frac{1}{2}ba\sin C$$ |

The Law of Sines for Any Triangle

The three areas for the scalene or the obtuse triangle can be equated.

$$\frac{1}{2}ca\sin B = \frac{1}{2}cb\sin A = \frac{1}{2}ba\sin C$$

Multiplying by $\dfrac{2}{abc}$ yields the Law of Sines.   $\boxed{\dfrac{\sin B}{b} = \dfrac{\sin A}{a} = \dfrac{\sin C}{c}}$

The Law of Sines shows that, if a triangle exists, three of its six parts must be known.

The Law of Cosines is derived in the following table.

| Deriving the Law of Cosines | |
|---|---|
| 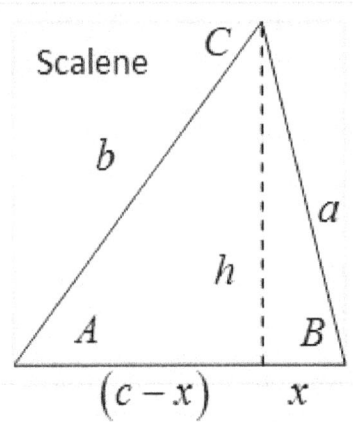 Scalene | • $h$ is height, angles are A, B and C. <br> • Sides are $a$, $b$ and $c$. <br> • $c$ is segmented into $(c - x)$ and $x$. <br><br> • <u>For angle A</u> $\cos A = \dfrac{c-x}{b} \rightarrow x = c - b * \cos A$ <br><br> $\sin A = \dfrac{h}{b} \rightarrow h = b * \sin A$ <br><br> • <u>For side $a$</u> $a^2 = x^2 + h^2 = (c - b*\cos A)^2 + (b*\sin A)^2$ <br><br> $\boxed{a^2 = b^2 + c^2 - 2b * c * \cos A}$ |
| 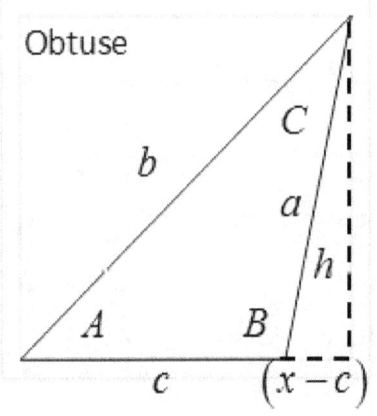 Obtuse | • Side $c$ is augmented as shown. <br><br> • <u>For angle A</u> $\cos A = \dfrac{c+x-c}{b} \rightarrow x = b * \cos A$ <br><br> $\sin A = \dfrac{h}{b} \rightarrow h = b * \sin A$ <br><br> • <u>For side $a$</u> $a^2 = (x-c)^2 + h^2 = (b*\cos A - c)^2 + (b*\sin A)^2$ <br><br> $\boxed{a^2 = b^2 + c^2 - 2b * c * \cos A}$ |

*The* Law of Cosines for Any Triangle

• For angle $A$ and side $a$ $\boxed{a^2 = b^2 + c^2 - 2b * c * \cos A}$

Similarly,

• For angle $B$ and side $b$ $\boxed{b^2 = a^2 + c^2 - 2a * c * \cos B}$

• For angle $C$ and side $c$ $\boxed{c^2 = a^2 + b^2 - 2a * b * \cos C}$

Note: If either $A$, $B$ or $C$ is a right angle, the Law of Cosines reverts to the Pythagorean theorem.

The Law of Cosines shows that, if a triangle exists, three of its six parts must be known.

The following table shows use of the **Law of Sines** to solve Type 1 triangles.

| Type 1: Given Two Angles and a Side opposite One of Them (Referred to as AAS) |
|---|
| ( Note: Type 1 always yields a unique triangle.) |

Given two angles ($\alpha$ = 40 and $\gamma$ = 80) and an opposite side ($c$ = 10).

Compute the third angle $\beta$ = 180 – 40 – 80 = 60.

Draw a sketch.

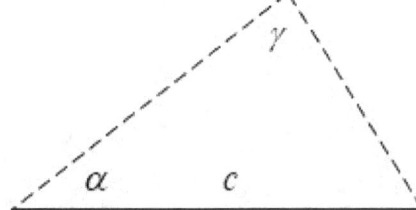

Compute the remaining two sides.

- Using the Law of Sines $\dfrac{c}{\sin\gamma} = \dfrac{a}{\sin\alpha}$ → $a = \dfrac{c}{\sin\gamma}\sin\alpha = 6.53$

- Using the Law of Sines $\dfrac{c}{\sin\gamma} = \dfrac{b}{\sin\beta}$ → $b = \dfrac{c}{\sin\gamma}\sin\beta = 8.8$

Given two angles ($\alpha$ = 120 and $\gamma$ = 40) and an opposite side ($a$ = 10).

Compute the third angle $\beta$ = 180 – 120 – 40 = 20

Draw a sketch.

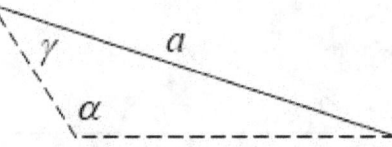

Compute the remaining two sides.

- Using the Law of Sines $\dfrac{a}{\sin\alpha} = \dfrac{b}{\sin\beta}$ → $b = \dfrac{a}{\sin\alpha}\sin\beta = 3.95$

- Using the Law of Sines $\dfrac{a}{\sin\alpha} = \dfrac{c}{\sin\gamma}$ → $c = \dfrac{a}{\sin\alpha}\sin\gamma = 7.42$

The following table shows use of the **Law of Sines** to solve Type 2 triangles.

| Type 2: Given Two Angles and the Side between Them (Referred to as ASA) |
|---|
| ( Note: Type 2 always yields a unique triangle.) |

Given two angles ($\beta$ = 60 and $\gamma$ = 80) and the side between them ($a$ = 6.53).

Compute the third angle $\alpha$ = 180 − 60 − 80 = 40.

Draw a sketch.

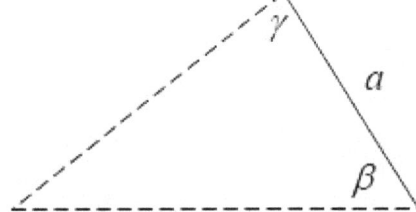

Compute the remaining two sides.

- Using the Law of Sines $\dfrac{a}{\sin\alpha} = \dfrac{b}{\sin\beta}$ → $b = \dfrac{a}{\sin\alpha}\sin\beta = 8.8$

- Using the Law of Sines $\dfrac{a}{\sin\alpha} = \dfrac{c}{\sin\gamma}$ → $c = \dfrac{a}{\sin\alpha}\sin\gamma = 10$

Given two angles ($\alpha$ = 120 and $\beta$ = 20) and the side between them ($c$ = 7.42).

Compute the third angle $\gamma$ = 180 − 120 − 20 = 40.

Draw a sketch.

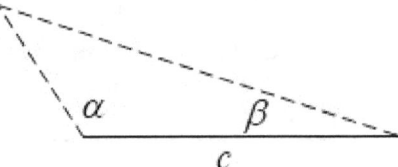

Compute the remaining two sides.

- Using the Law of Sines $\dfrac{a}{\sin\alpha} = \dfrac{c}{\sin\gamma}$ → $a = \dfrac{c}{\sin\gamma}\sin\alpha = 10$

- Using the Law of Sines $\dfrac{b}{\sin\beta} = \dfrac{c}{\sin\gamma}$ → $b = \dfrac{c}{\sin\gamma}\sin\beta = 3.95$

Type 3 Triangles: These are when the *givens* are two sides and the angle opposite one of them. This is commonly referred to as SSA. Specifications of this type do not always yield a unique triangle. As can be seen in the following figure, side *a* is dangling. Before calculations, it's difficult to know where side *a* will be. There may no solution, two solutions, or one solution.

| Type 3: Given Two Sides and an Angle opposite One of Them (Referred to as SSA) | | |
|---|---|---|
| Side *a* is too short and there is no solution. | The length of *a* allows two possible solutions. | As the length of *a* increases, there is only one solution. |
| *b*   *a*   α | *b*   *a*   α    A right triangle is also possible. | *b*   *a*   α |

The following five examples show the application of the **Law of Sines** to solving Type 3 triangles.

| Type 3: Example 1 |
|---|
| Given two sides (*a* = 10 and *b* = 8.8) and an opposite angle (α = 40). |

- The Law of Sines for an angle $\sin\beta = \dfrac{\sin\alpha}{a} b = 0.565$. There are two possible solutions.
$$\beta = 34.4 \text{ and } \beta = 145.6.$$

- If $\beta = 34.4$, then the remaining angle is $\gamma = 180 - \alpha - \beta = 105.6$.
  If $\beta = 145.6$, then the remaining angle is $\gamma = 180 - \alpha - \beta = -5.6$, which is not possible.
  The conclusion is that $\beta = 34.4$ and $\gamma = 105.6$.

- The Law of Sines for the third side $c = \dfrac{a}{\sin\alpha}\sin\gamma = 14.98$.

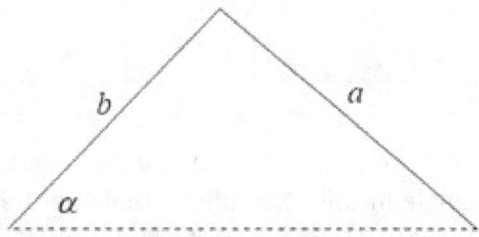

## Type 3: Example 2

Given two sides ($a$ = 10 and $b$ = 3.95) and an opposite angle ($\alpha$ = 120).

- The Law of Sines for another angle yields $\sin\beta = \dfrac{\sin\alpha}{a}b = 0.342$. $\beta$ can be 20 or 160.

- If $\beta = 20$, then the remaining angle is $\gamma = 180 - \alpha - \beta = 40$.
  If $\beta = 160$, then the remaining angle is $\gamma = 180 - \alpha - \beta = -100$, which is not possible.
  The conclusion is that $\beta = 20$ and $\gamma = 40$.

- From the Law of Sines, $c = \dfrac{a}{\sin\alpha}\sin\gamma = 7.42$.

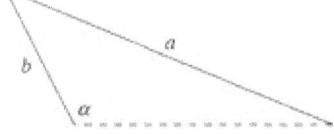

## Type 3: Example 3

Given two sides ($a$ = 6.53 and $b$ = 8.8) and an opposite angle ($\alpha$ = 40).

- The Law of Sines for another angle yields $\sin\beta = \dfrac{\sin\alpha}{a}b = 0.866$. $\beta$ can be 60 or 120.

- If $\beta = 60$, then the remaining angle is $\gamma = 180 - \alpha - \beta = 80$.
  If $\beta = 120$, then the remaining angle is $\gamma = 180 - \alpha - \beta = 20$.
  The conclusion is that there are two possible triangles.

$$\left(\beta = 60, \gamma = 80\right) \text{ and } \left(\beta = 120, \gamma = 20\right)$$

| | |
|---|---|
| If $\left(\beta = 60, \gamma = 80\right)$, then the Law of Sines for the third side: $c = \dfrac{a}{\sin\alpha}\sin\gamma = 10$. | If $\left(\beta = 120, \gamma = 20\right)$, then the Law of Sines for the third side: $c = \dfrac{a}{\sin\alpha}\sin\gamma = 3.47$. |
| 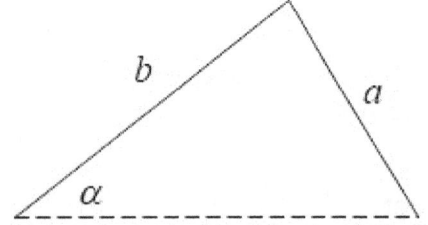 | 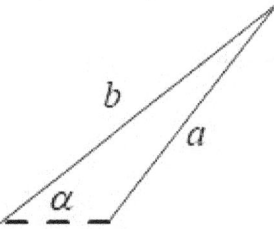 |

The scalene triangle on the left is the same as the one in the examples of Types 1, 2, 4 and 5. The reason for the second triangle in this example, is that there are two possibilities for the angle $\gamma$. This clearly shows why Type 3 is not a good way to specify a triangle.

---

## Type 3: Example 4

Given two sides ($a$ = 4 and $b$ = 8.8) and an opposite angle ($\alpha$ = 40).

• The Law of Sines for an angle: $\sin\beta = \dfrac{\sin\alpha}{a}b = 1.41$. Because $\sin\beta$ cannot exceed unity, this example is not a triangle. As shown in the following figure, side $a$ is too short.

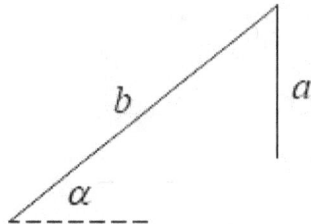

---

## Type 3: Example 5

Given two sides ($a$ = 3.95 and $b$ = 3.9) and an opposite angle ($\alpha$ = 120).

• The Law of Sines for an angle: $\sin\beta = \dfrac{\sin\alpha}{a}b \approx 0.866$. There are two possible solutions.

$$\beta = 120 \text{ and } \beta = 60.$$

• If $\beta = 120$, then the remaining angle is $\gamma = 180 - \alpha - \beta = -60$. This is not possible.
 If $\beta = 60$, then the remaining angle is $\gamma = 180 - \alpha - \beta = 0$, which means there is no triangle. This is shown in the following figure.

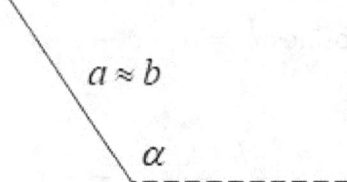

The following table shows use of the **Law of Cosines** to solve Type 4 triangles.

---

### Type 4: Given Two Sides and the Angle between Them (Referred to as SAS)
( Note: Type 4 always yields a unique triangle.)

---

Given two sides ($b$ = 8.8 and $c$ = 10) and the angle between them ($\alpha$ = 40).

Draw a sketch.

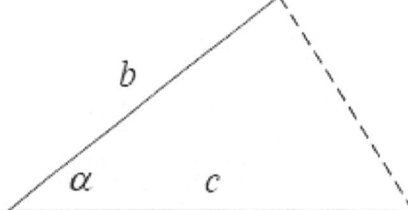

• The third side from the Law of Cosines $a^2 = b^2 + c^2 - 2bc\cos\alpha \quad \rightarrow \quad a = 6.53$

• Another angle from the Law of Cosines

$$\cos\beta = \frac{b^2 - a^2 - c^2}{(-2ac)} \quad \rightarrow \quad \beta = 60$$

• The remaining angle $\gamma = 180 - \alpha - \beta = 80$.

---

Given two sides ($a$ = 10 and $c$ = 7.42) and the angle between them ($\beta$ = 20).

Draw a sketch.

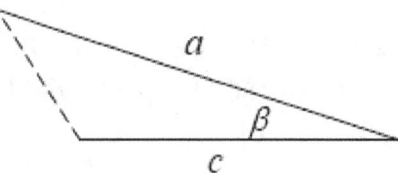

• The third side from the Law of Cosines $b^2 = a^2 + c^2 - 2ac\cos\beta \quad \rightarrow \quad b = 3.95$

• Another angle from the Law of Cosines

$$\cos\alpha = \frac{a^2 - b^2 - c^2}{(-2bc)} \quad \rightarrow \quad \alpha = 120$$

• The remaining angle $\gamma = 180 - \alpha - \beta = 40$.

The following table shows use of the **Law of Cosines** to solve Type 5 triangles.

| Type 5: Given Three Sides (Referred to as SSS) |
| :---: |
| ( Note: Type 5 always yields a unique triangle.) |

Given three sides ($a$ = 6.53 and $b$ = 8.8 and $c$ = 10 ).

Draw a sketch.

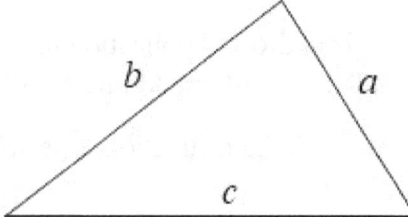

- An angle from the Law of Cosines

$$\cos\alpha = \frac{a^2 - b^2 - c^2}{(-2bc)} \quad \rightarrow \quad \alpha = 40$$

- Another angle from the Law of Cosines

$$\cos\beta = \frac{b^2 - a^2 - c^2}{(-2ac)} \quad \rightarrow \quad \beta = 60$$

- The remaining angle $\gamma$ = 180 - $\alpha$ − $\beta$ = 80.

Given three sides ($a$ = 10, $b$ =3.95 and $c$ = 7.42).

Draw a sketch.

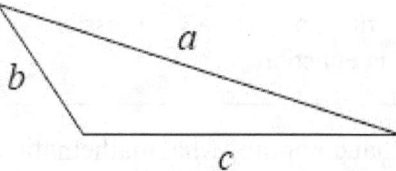

- An angle from the Law of Cosines

$$\cos\alpha = \frac{a^2 - b^2 - c^2}{(-2bc)} \quad \rightarrow \quad \alpha = 120$$

- Another angle from the Law of Cosines

$$\cos\beta = \frac{b^2 - a^2 - c^2}{(-2ac)} \quad \rightarrow \quad \beta = 20$$

- The remaining angle $\gamma$ = 180 - $\alpha$ − $\beta$ = 40.

# Trigonometry – Programming in Chapter 6

Another feature of the VBA language must be presented in order to do the programs in this chapter. We have used VBA functions like these:

- the square root, Sqr(x);
- the absolute value, Abs(x).

We have also used Excel functions like these:

- Round-off, Application.Round(x, n);
- The logarithm, Application.Log(x, n).

There is a third type. It's called the user-defined function. The following table shows the syntax for this.

| Syntax for a User-Defined Function |
|---|
| Sub main() <br> • <br> • <br>   y = function_name(argument list) <br> • <br> • <br> End Sub <br><br> Function function_name(argument list) <br> • <br> • <br>   function_name = expression <br> End Function |

Let's write a user-defined function that computes what mathematicians call *n factorial*. This is a product of non-negative integers. It occurs so frequently that it has its own symbol ( n! ). By definition,

$$n! = \prod_{k=1}^{n} k = n*(n-1)*(n-2)*(n-3)* \bullet\bullet\bullet *3*2*1,$$

and

$$0! = 1.$$

For example, $(2n-1)! = (2n-1)*(2n-2)*(2n-3)*(2n-4)* \bullet\bullet\bullet *3*2*1.$

The following table shows the program.

| Example nfactorial | |
|---|---|
| Code | Syntax |
| ```Sub main()    n = Cells(1, 1)    X = nfactorial(n)    Cells(1, 2) = X End Sub  Function nfactorial(n)    nfactorial = 1    For i = 1 To n       nfactorial = nfactorial * i    Next i End Function``` | • If n = 0, the statements in the For…Next loop are not executed. That achieves the result<br><br>$$0 ! = 1 ! = 1$$<br><br>• The equal sign really means *is replaced by*. This is shown by the statement nfactorial = nfactorial * i |

In this chapter, we showed the following:

- The sine function;
- The cosine function;
- The tangent function;
- The inverse sine function;
- The inverse cosine function;
- The inverse tangent function.

The following six tables show the programs that created the plots of each, along with its Taylor series representation. In the programs, the Taylor series and the **n factorials** are computed by user-defined functions.

## The Sine Function: Program Comparing Taylor Series with the VBA Function

```
Sub TSsine()
  Dx = 30 * 4 * Atn(1) / 180: rad_deg = 180 / (4 * Atn(1))
  x = 0
  For i = 1 To 13
    TS_sin = sine(x)
    XL_sin = Sin(x)
    Cells(i, 1) = x * rad_deg
    Cells(i, 2) = Application.Round(TS_sin, 4)
    Cells(i, 3) = Application.Round(XL_sin, 4)
    x = x + Dx
  Next i
End Sub

Function sine(x)
  sine = 0
  For N = 1 To 10
    nm1 = N - 1
    pow = 2 * N - 1
    nf = nfactorial(pow)
    sine = sine + (-1) ^ nm1 * (x ^ pow) / nf
  Next N
End Function

Function nfactorial(N)
  nfactorial = 1
  For i = 1 To N
    nfactorial = nfactorial * i
  Next i
End Function
```

- Taylor series: sine(x)
- VBA:    Sin(x)
- x Range: 0 to $360^0$

- $sine(x) = \sum_{n=1}^{10}(-1)^{n-1}\dfrac{x^{2n-1}}{(2n-1)!}$
  - Valid for all x.
  - $(2n-1)! = (2n-1)*(2n-2)* \bullet\bullet\bullet *3*2*1$

- n factorial

The following table is the output.

| Angle | Series 1 | Series 2 |
|---|---|---|
| 0 | 0 | 0 |
| 30 | 0.5 | 0.5 |
| 60 | 0.866 | 0.866 |
| 90 | 1 | 1 |
| 120 | 0.866 | 0.866 |
| 150 | 0.5 | 0.5 |
| 180 | 0 | 0 |
| 210 | -0.5 | -0.5 |
| 240 | -0.866 | -0.866 |
| 270 | -1 | -1 |
| 300 | -0.866 | -0.866 |
| 330 | -0.5002 | -0.5 |
| 360 | -0.001 | 0 |

Series 1: Taylor Series
Series 2: VBA Sin(x)

○ Series1
— Series2

## The Cosine Function: Program Comparing Taylor Series with the VBA Function

```
Sub TScosine()
  Dx = 30 * 4 * Atn(1) / 180: rad_deg = 180 / (4 * Atn(1))
  x = 0
  For i = 1 To 13
    TS_cos = cosine(x)
    XL_cos = Cos(x)
    Cells(i, 1) = x * rad_deg
    Cells(i, 2) = Application.Round(TS_cos, 4)
    Cells(i, 3) = Application.Round(XL_cos, 4)
    x = x + Dx
  Next i
End Sub

Function cosine(x)
    cosine = 0
    For N = 0 To 10
      pow = 2 * N
      nf = nfactorial(pow)
      cosine = cosine + (-1) ^ N * (x ^ pow) / nf
    Next N
End Function

Function nfactorial(N)
    nfactorial = 1
  For i = 1 To N
    nfactorial = nfactorial * i
  Next i
End Function
```

- Taylor series: cosine(x)
- VBA:   Cos(x)
- x Range: 0 to $360^0$

- $\cosine(x) = \sum_{n=0}^{10} (-1)^n \dfrac{x^{2n}}{(2n)!}$

  - Valid for all x.
  - $(2n)! = (2n)*(2n-1)* \bullet\bullet\bullet *3*2*1$

- n factorial

The following table is the output.

| Angle | Series 1 | Series 2 | |
|-------|----------|----------|---|
| 0 | 1 | 1 | |
| 30 | 0.866 | 0.866 | |
| 60 | 0.5 | 0.5 | |
| 90 | 0 | 0 | |
| 120 | -0.5 | -0.5 | |
| 150 | -0.866 | -0.866 | |
| 180 | -1 | -1 | |
| 210 | -0.866 | -0.866 | |
| 240 | -0.5 | -0.5 | |
| 270 | 0 | 0 | |
| 300 | 0.5 | 0.5 | |
| 330 | 0.8661 | 0.866 | |
| 360 | 1.0003 | 1 | |

Series 1: Taylor Series
Series 2: VBA Cos(x)

○   Series1
—   Series2

## The Tangent Function: Program Comparing Taylor Series with the VBA Function

```
Sub TStangent()
  Dx = 5 * 4 * Atn(1) / 180: rad_deg = 180 / (4 * Atn(1))
  x = -85 * 4 * Atn(1) / 180
  For i = 1 To 35
    TS_tan = tangent(x)
    XL_tan = Tan(x)
    Cells(i, 1) = x * rad_deg
    Cells(i, 2) = Application.Round(TS_tan, 4)
    Cells(i, 3) = Application.Round(XL_tan, 4)
    x = x + Dx
  Next i
End Sub
Function tangent(x)
  sinX = 0
  For N = 1 To 10
    nm1 = N – 1: pow = 2 * N - 1
    nf = nfactorial(pow)
    sinX = sinX + (-1) ^ nm1 * (x ^ pow) / nf
  Next N
  cosinX = 0
  For N = 0 To 10
    pow = 2 * N:  nf = nfactorial(pow)
    cosinX = cosinX + (-1) ^ N * (x ^ pow) / nf
  Next N
    tangent = sinX / cosinX
End Function
Function nfactorial(N)
   nfactorial = 1
 For i = 1 To N
   nfactorial = nfactorial * i
 Next i
End Function
```

- Taylor series: tangent(x)
- VBA:  Tan(x)
- x Range: -85 to +85°

- Taylor series computed from the ratio:
  - $tangent(x) = \dfrac{sine(x)}{cosine(x)}$
  - valid for all x

- n factorial

The following table is its output.

| Angle | Series 1 | Series 2 | |
|-------|----------|----------|---|
| -85 | -11.4301 | -11.4301 | |
| -80 | -5.6713 | -5.6713 | |
| -75 | -3.7321 | -3.7321 | |
| -70 | -2.7475 | -2.7475 | |
| -65 | -2.1445 | -2.1445 | |
| -60 | -1.7321 | -1.7321 | |
| -55 | -1.4281 | -1.4281 | |
| -50 | -1.1918 | -1.1918 | |
| -45 | -1 | -1 | |
| -30 | -0.5774 | -0.5774 | |
| -20 | -0.364 | -0.364 | |
| -10 | -0.1763 | -0.1763 | |
| 0 | 0 | 0 | |

Series 1: Taylor series
Series 2: VBA Tan(x)

○  Series1
——  Series2

## The Inverse Sine Function: Program Comparing Taylor Series with the VBA Function

```
Sub TSasine()
 Dim s(100)
  rad_deg = 180 / (4 * Atn(1))
  s(1) = -1
  For i = 2 To 21
   s(i) = s(i - 1) + 0.1
  Next i
  s(1) = -0.99: s(21) = 0.99
  For i = 1 To 21
   x = s(i)
   TS_asine = asine(x) * rad_deg
   XL_asine = Application.Asin(x) * rad_deg
   Cells(i, 1) = x
   Cells(i, 2) = Application.Round(TS_asine, 4)
   Cells(i, 3) = Application.Round(XL_asine, 4)
  Next i
End Sub

Function asine(x)
   asine = 0
   For N = 0 To 10
    twoN = 2 * N: twoT2n = 2 ^ twoN: tnP1 = 2 * N + 1
    tn_f = nfactorial(twoN): n_f = nfactorial(N)
    asine = asine + (tn_f * x ^ tnP1) / (twoT2n * n_f ^ 2 * tnP1)
   Next N
End Function

Function nfactorial(N)
   nfactorial = 1
  For i = 1 To N
   nfactorial = nfactorial * i
  Next i
End Function
```

- Note: the minimum and maximum values for x

- Taylor series: asine(x)
- VBA:   Asin(x)
- x range: -0.99 to 0.99

- $\text{asine}(x) = \sum_{n=0}^{10} \dfrac{(2n)!}{2^{2n} \, (n!)^2} \dfrac{x^{2n+1}}{(2n+1)}$

  - $(2n)! = (2n)*(2n-1)* \bullet\bullet\bullet *3*2*1$
  - $(n!)^2 = \{(n)*(n-1)*(n-2)* \bullet\bullet\bullet *3*2*1\}^2$
  - valid for $-0.99 < x < +0.99$

- n factorial

The following table is its output.

| X input | Series 1 | Series 2 | |
|---|---|---|---|
| -0.99 | -78.1669 | -81.8904 | |
| -0.9 | -64.0211 | -64.1581 | |
| -0.8 | -53.1243 | -53.1301 | |
| -0.7 | -44.4268 | -44.427 | |
| -0.6 | -36.8699 | -36.8699 | |
| -0.5 | -30 | -30 | |
| -0.4 | -23.5782 | -23.5782 | |
| -0.3 | -17.4576 | -17.4576 | |
| -0.2 | -11.537 | -11.537 | |
| -0.1 | -5.7392 | -5.7392 | |
| 0 | 0 | 0 | |

Series 1: Taylor Series
Series 2: VBA Asin(x)

○ Series1
— Series2

## The Inverse Cosine Function: Program Comparing Taylor Series with the VBA Function

```
Sub TSacosine()
 Dim s(100)
 rad_deg = 180 / (4 * Atn(1))
 s(1) = -1
 For i = 2 To 21
   s(i) = s(i - 1) + 0.1
 Next i
 s(1) = -0.99: s(21) = 0.99
 For i = 1 To 21
   x = s(i)
   TS_acosine = acosine(x) * rad_deg
   XL_acosine = Application.Acos(x) * rad_deg
   Cells(i, 1) = x
   Cells(i, 2) = Application.Round(TS_acosine, 4)
   Cells(i, 3) = Application.Round(XL_acosine, 4)
 Next i
End Sub
Function acosine(x)
   pie = 4 * Atn(1)
   asinX = 0
   For N = 0 To 10
     twoN = 2 * N: twoT2n = 2 ^ twoN: tnP1 = 2 * N + 1
     tn_f = nfactorial(twoN): n_f = nfactorial(N)
     asinX = asinX + (tn_f * x ^ tnP1) / (twoT2n * n_f ^ 2 * tnP1)
     acosine = (pie / 2) - asinX
   Next N
End Function
Function nfactorial(N)
   nfactorial = 1
 For i = 1 To N
   nfactorial = nfactorial * i
 Next i
End Function
```

- Note: the minimum and maximum values for x.

- Taylor series: acosine(x)
- VBA:  Acos(x)
- x Range: -0.99 to 0.99

- acosine(x) $= \displaystyle\sum_{n=0}^{10}\left( \frac{\pi}{2} - \frac{(2n)!}{2^{2n}\,(n!)^2}\frac{x^{2n+1}}{(2n+1)} \right)$

  - $(2n)! = (2n)*(2n-1)* \bullet\bullet\bullet *3*2*1$
  - $(n!)^2 = \{(n)*(n-1)*(n-2)* \bullet\bullet\bullet *3*2*1\}^2$
  - Valid:  $-0.99 < x < 0.99$.

- n factorial

The following table is its output.

| X Input | Series 1 | Series 2 |
|---:|---:|---:|
| -0.99 | 168.1669 | 171.8904 |
| -0.9 | 154.0211 | 154.1581 |
| -0.8 | 143.1243 | 143.1301 |
| -0.7 | 134.4268 | 134.427 |
| -0.6 | 126.8699 | 126.8699 |
| -0.5 | 120 | 120 |
| -0.4 | 113.5782 | 113.5782 |
| -0.3 | 107.4576 | 107.4576 |
| -0.2 | 101.537 | 101.537 |
| -0.1 | 95.7392 | 95.7392 |
| 0 | 90 | 90 |

Series 1: Taylor Series
Series 2: VBA Acos(x)

○  Series1
——  Series2

## The Inverse Tangent Function: Program Comparing Taylor Series with the VBA Function

```
Sub TSatangent()
 Dim s(100)
  rad_deg = 180 / (4 * Atn(1))
  For i = 1 To 35
    s(i) = Cells(i, 1)
  Next i
  For i = 1 To 35
    x = s(i)
    TS_atan = atangent(x) * rad_deg
    XL_atan = Atn(x) * rad_deg
    Cells(i, 1) = x
    Cells(i, 2) = Application.Round(TS_atan, 4)
    Cells(i, 3) = Application.Round(XL_atan, 4)
  Next i
End Sub
Function atangent(x)
  pie = 4 * Atn(1)
  sum1 = 0: sum2 = 0
  For N = 0 To 3
    M1toN = (-1) ^ N
    Tnp1 = 2 * N + 1
    sum1 = sum1 + M1toN / (Tnp1 * x ^ Tnp1)
    sum2 = sum2 + M1toN * x ^ Tnp1 / Tnp1
  Next N
  If x < -1 Then
    atangent = -pie / 2 - sum1
  ElseIf x > 1 Then
    atangent = pie / 2 - sum1
  Else
    atangent = sum2
  End If
  If x = 1 Then atangent = pie / 4
  If x = -1 Then atangent = -pie / 4
End Function
```

- Values for x are the output of tangent(x)

- Taylor series: atangent(x)
- VBA:   Atn(x)
- x Range: from tangent(x) [-85 to 85$^0$]

- If $x < -1$, then

$$\text{atangent(x)} = -\frac{\pi}{2} - \sum_{n=0}^{\infty}(-1)^n \frac{1}{(2n+1)x^{(2n+1)}}$$

- If $-1 < x < 1$, then

$$\text{atangent(x)} = \sum_{n=0}^{\infty}(-1)^n \frac{x^{(2n+1)}}{(2n+1)}$$

- If $x > 1$, then

$$\text{atangent(x)} = \frac{\pi}{2} - \sum_{n=0}^{\infty}(-1)^n \frac{1}{(2n+1)x^{(2n+1)}}$$

- If $x = 1$, then $\text{atangent(x)} = \frac{\pi}{4}$

- If $x = -1$, then $\text{atangent(x)} = -\frac{\pi}{4}$

The following table is its output.

| X Input | Series 1 | Series 2 |
|---|---|---|
| -11.4301 | -85 | -85 |
| -5.6713 | -80 | -80 |
| -3.7321 | -75.0002 | -75.0002 |
| -2.7475 | -70.0008 | -70.0002 |
| -2.1445 | -65.0056 | -64.9999 |
| -1.7321 | -60.0364 | -60.0007 |
| -1.4281 | -55.1835 | -54.9991 |
| -1.1918 | -50.838 | -50.0011 |
| -1 | -45 | -45 |
| -0.5774 | -29.9664 | -30.0021 |
| -0.364 | -20.0009 | -20.0015 |
| -0.1763 | -9.9985 | -9.9985 |
| 0 | 0 | 0 |

Series 1: Taylor Series
Series 2: VBA Atn(x)

# Chapter 7: The Conics

Conics have been studied for about 2500 years. Conic sections are curves that are created when straight cuts are made through right circular cones. If a cut is perpendicular to a cone's axis, a circle is made. Other cuts produce a parabola or an ellipse. If a cut goes through two cones that are lined up with their tips touching, a hyperbola is made.

This chapter is divided into the following sections:

- Defining the conics in rectangular and polar coordinates;
- Translating the conics;
- Rotating the conics;
- Unrotating the conics;
- Discriminants for conic equations;
- Interesting cases;
- Programming the math used in this chapter.

# Parabola

The equation of a parabola is $y^2 = Kx$. To plot it, we solve for y: $y = \pm\sqrt{Kx}$. If $K$ is positive,

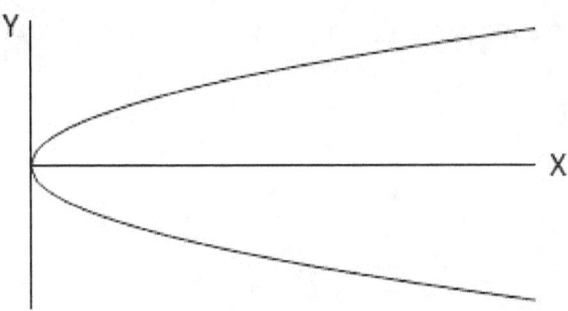

Let's put a geometric interpretation on $K$. Our parabola, $y^2 = Kx$, is formally defined by its focus and its directrix. Let's say that its focus is located at x = p. Its directrix, which is a line perpendicular to the x-axis, will then be located at x = −p. This is shown in the following figure.

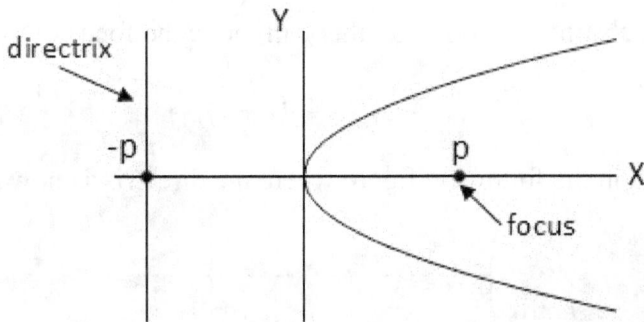

The following figure adds two distances, $d_1$ and $d_2$.

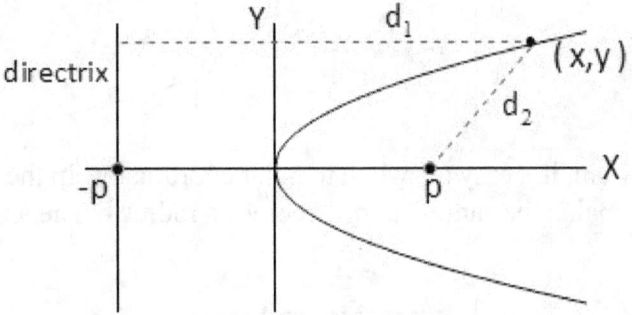

We can now formally define a parabola. At each point on a parabola, $d_1 = d_2$. At the point $(x, y)$,

$d_1 = p + x$ and $d_2 = \sqrt{(x-p)^2 + (y)^2}$. Since $d_1^2 = d_2^2$,

$$(p+x)^2 = (x-p)^2 + (y)^2.$$
$$p^2 + 2px + x^2 = x^2 - 2px + p^2 + y^2$$

This yields the equation for a horizontal parabola.

$$y^2 = 4px$$

Equation P.1

The following figure shows the parabola for p = 1/2.

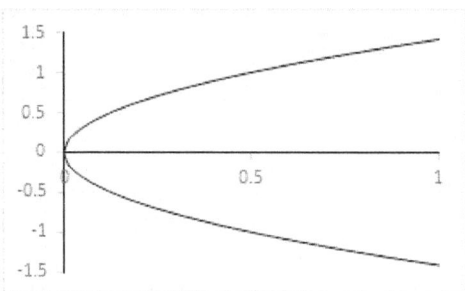

If we are given the plot, it's easy to compute the value of p. At a distance x from the vertex, pick off the value of y. Compute $p = y^2/(4x)$.

In equation P.1, if we substitute $(x+p) = x$, this will move the focus of the parabola to the origin.

$$y^2 = 4p(x+p)$$

Equation P.2

This equation is plotted in the following figure where the directrix is now at x = −2p.

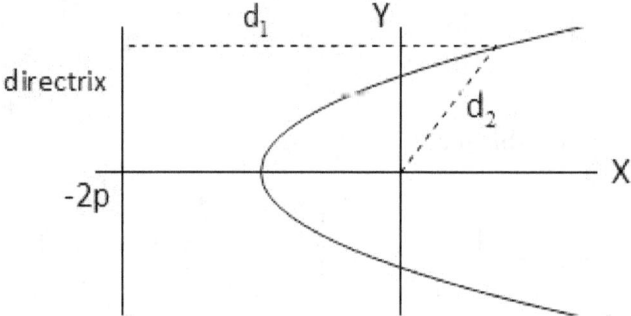

With the focus at the origin, it's easy to switch to polar coordinates. In the following figure, the distance to the directrix ( $d_1$ ) remains the same, but $d_2$ becomes a radius r. The angle $\theta$ is also introduced.

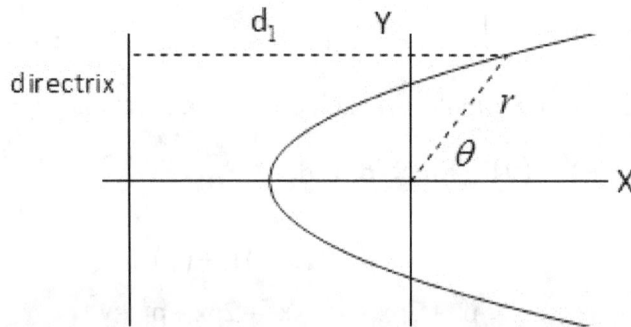

The equation for $d_1$ is $d_1 = 2p+x$. As before, $d_1 = d_2 = r$. Therefore, $r = 2p + x = 2p + r\cos\theta$.

This yields the polar equation for a horizontal parabola.

$$r = \frac{2p}{1 - \cos\theta}$$

Equation P.3

Equations P.2 and P.3 describe the same parabola. To show this, let's transform equation P.3 back into rectangular coordinates. Since $r = 2p + x$, $r^2 = x^2 + y^2 = (2p + x)^2 = 4p^2 + 4px + x^2$. This reduces to $y^2 = 4p(x + p)$, which is equation P.2.

Using **p** = 1/2, Equations P.2 and P.3 produce the parabola shown in the following table.

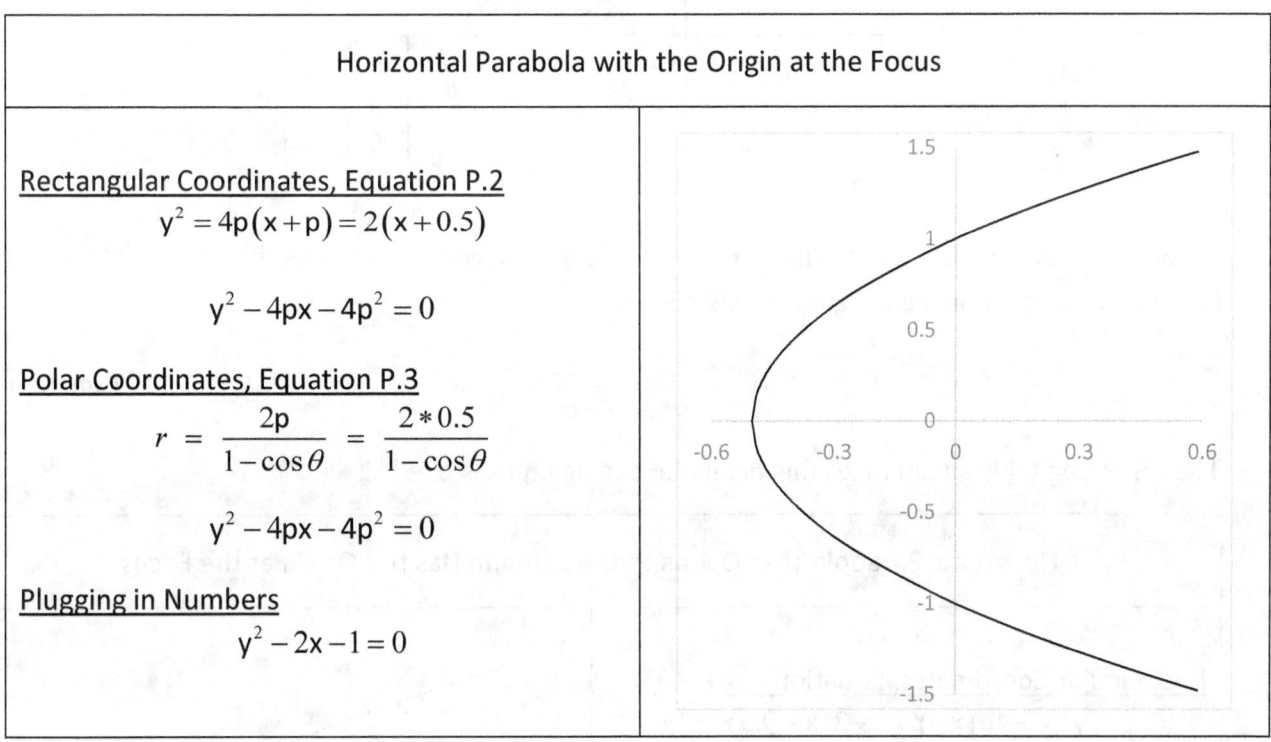

| Horizontal Parabola with the Origin at the Focus |
|---|

Rectangular Coordinates, Equation P.2
$$y^2 = 4p(x + p) = 2(x + 0.5)$$

$$y^2 - 4px - 4p^2 = 0$$

Polar Coordinates, Equation P.3
$$r = \frac{2p}{1 - \cos\theta} = \frac{2*0.5}{1 - \cos\theta}$$

$$y^2 - 4px - 4p^2 = 0$$

Plugging in Numbers
$$y^2 - 2x - 1 = 0$$

We started our study of the parabola with the following equation: $y^2 = Kx$, where $K$ was positive. What happens when $K$ is negative? For **y** to be a real number, **x** must be a negative number. The following is the plot for $K < 0$.

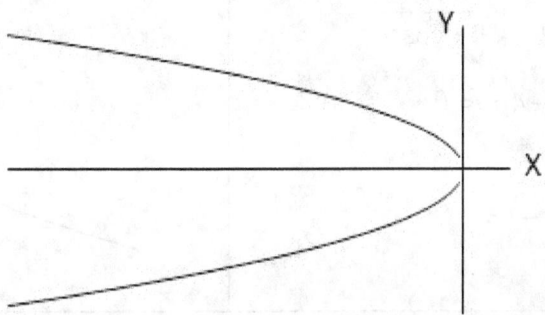

The equation for this parabola is $y^2 = -4px$. This allows p to remain a positive number. Let's move the focus of this parabola to the origin by using the following equation:

$$y^2 = -4p(x-p)$$
Equation P.4

The following figure defines the directrix and $d_1$ and $d_2$ for this horizontal parabola.

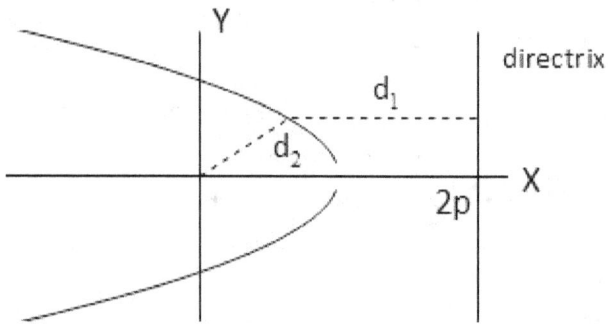

Here, $d_2 = r$ and $d_1 = 2p - x$, so that $r = 2p - x = 2p - r\cos\theta$. This yields the polar equation for the horizontal parabola on the negative x-axis.

$$r = \frac{2p}{1+\cos\theta}$$
Equation P.5

The following table summarizes this parabola, and again uses p = 1/2.

| Horizontal Parabola That Opens to the Left and Has the Origin at the Focus | |
| --- | --- |
| <u>Rectangular Coordinates, Equation P.4</u><br>$$y^2 = -4p(x-p) = -2(x-0.5)$$<br><br>$$y^2 + 4px - 4p^2 = 0$$<br><br><u>Polar Coordinates, Equation P.5</u><br>$$r = \frac{2p}{1+\cos\theta} = \frac{2*0.5}{1+\cos\theta}$$<br><br>$$y^2 + 4px - 4p^2 = 0$$<br><br><u>Plugging in Numbers</u><br>$$y^2 + 2x - 1 = 0$$ | 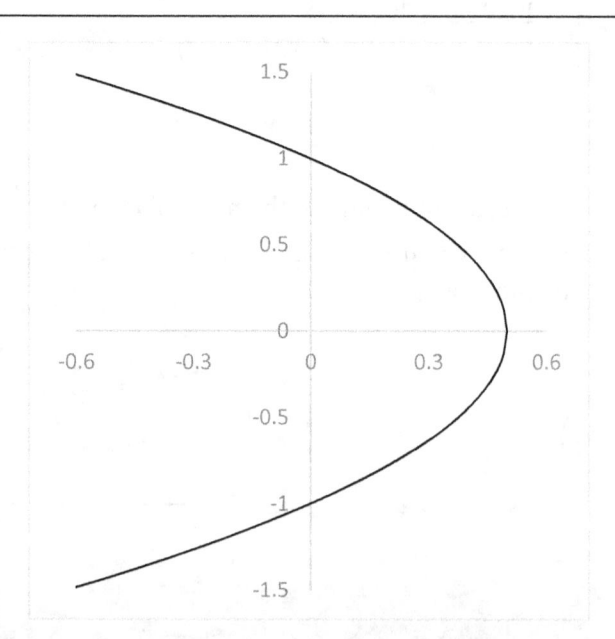 |

# Vertical Parabola

Horizontal parabolas are turned into vertical parabolas by reversing the roles of x and y. The horizontal axis will still be for the x variable. The following two tables show vertical parabolas.

| Vertical Parabola Opening Up | |
|---|---|
| **Rectangular Coordinates** $$x^2 = 4py = 2y$$ $$x^2 - 4py = 0$$ **Plugging in Numbers** $$x^2 - 2y = 0$$ | 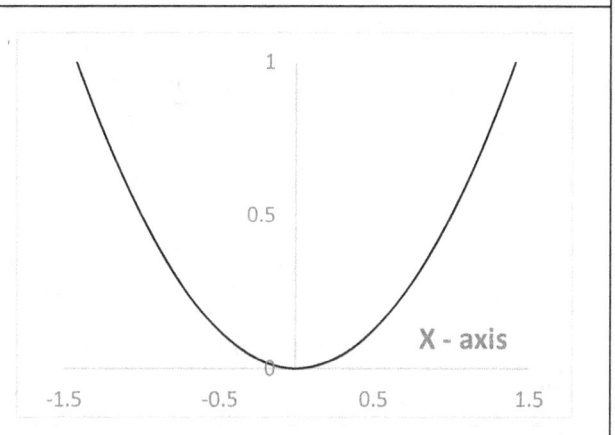 |

| Vertical Parabola Opening Down | |
|---|---|
| **Rectangular Coordinates** $$x^2 = -4py = -2y$$ $$x^2 + 4py = 0$$ **Plugging in Numbers** $$x^2 + 2y = 0$$ | 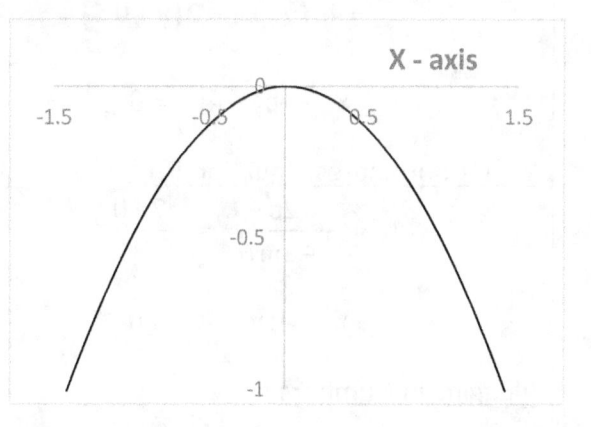 |

Let's derive the polar equation for the vertical parabola. First, for the parabola that opens upward, let's transfer the focus to the origin by substituting y = y + p.

$$x^2 = 4p(y+p)$$  Equation P.6

The following figure shows this parabola and its directrix at $y = -2p$.

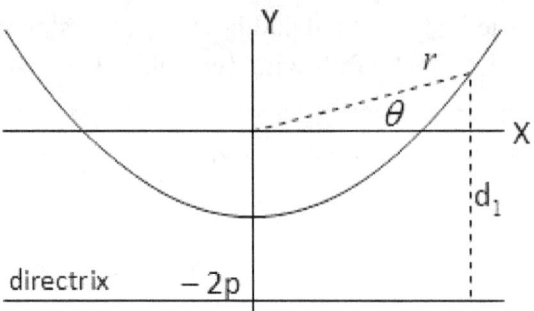

Here, $d_1 = y + 2p$ and $d_2 = r$. Therefore, $r = y + 2p = r\sin\theta + 2p$. This yields

$$r = \frac{2p}{1 - \sin\theta}$$
Equation P.7

The following table compares the rectangular with the polar equation, for a vertical parabola that opens upward.

| Vertical Parabola Opening Up, with the Focus at the Origin | |
| --- | --- |
| Rectangular Coordinates, Equation P.6<br>$x^2 = 4p(y+p) = 2(y+0.5)$<br><br>$x^2 - 4py - 4p^2 = 0$<br><br>Polar Coordinates, Equation P.7<br>$r = \dfrac{2p}{1-\sin\theta} = \dfrac{2*0.5}{1-\sin\theta}$<br><br>$x^2 - 4py - 4p^2 = 0$<br><br>Plugging in Numbers<br>$x^2 - 2y - 1 = 0$ | 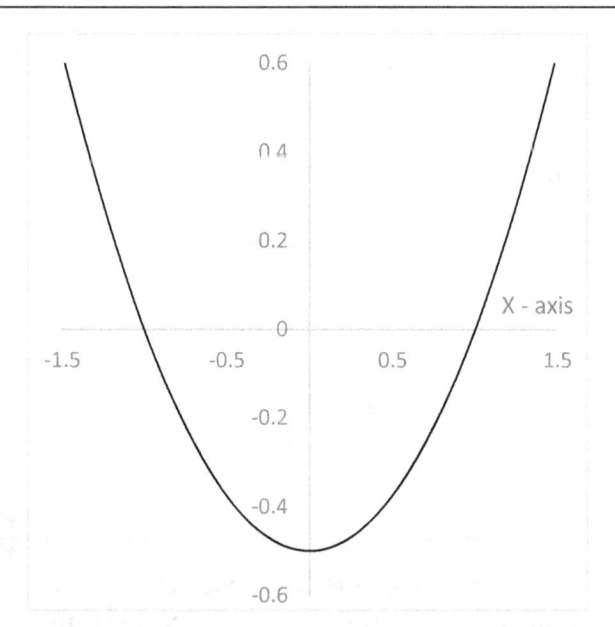 |

Now let's derive the polar equation for the vertical parabola that opens downward. As before, starting with the rectangular coordinates, $x^2 = -4py$, let's transfer the focus to the origin by substituting $y = y - p$.

$$x^2 = -4p(y-p)$$
Equation P.8

The following figure shows this parabola and its directrix at y = 2p.

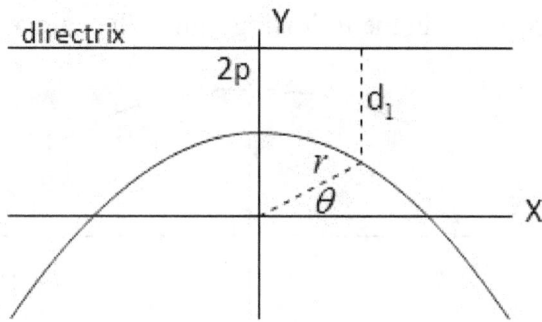

Here, $d_1 = 2p - y$ and $d_2 = r$. Therefore, $r = 2p - y = 2p - r\sin\theta$. This yields

$$r = \frac{2p}{1+\sin\theta}$$

Equation P.9

Equations P.8 and P.9 describe the same parabola. The following table summarizes these equations.

| Vertical Parabola Opening Down, with the Focus at the Origin | |
| --- | --- |
| Rectangular Coordinates, Equation P.8<br><br>$x^2 = -4p(y-p) = -2(y-0.5)$<br><br>$x^2 + 4py - 4p^2 = 0$<br><br>Polar Coordinates, Equation P.9<br><br>$r = \dfrac{2p}{1+\sin\theta} = \dfrac{2*0.5}{1+\sin\theta}$<br><br>$x^2 + 4py - 4p^2 = 0$<br><br>Plugging in Numbers<br><br>$x^2 + 2y - 1 = 0$ | (graph of parabola with X-axis, values -1.5, -0.5, 0.5, 1.5 on horizontal axis and 0.6, 0.4, 0.2, 0, -0.2, -0.4, -0.6 on vertical axis) |

We will now discuss the ellipse and the hyperbola.

# Ellipse

An ellipse has two foci. As shown in the following figure, they are at $\pm c$ .

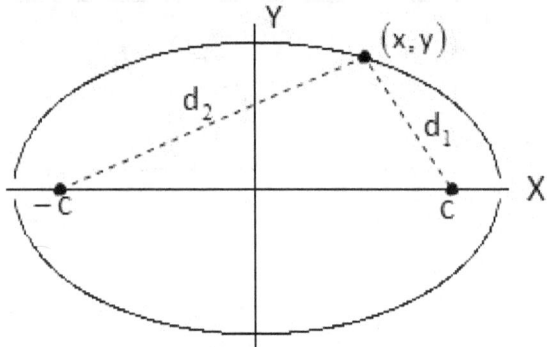

Also shown in the figure are two distances. $d_1$ is the distance from the focus at $c$ to a point $(x, y)$ on the ellipse. $d_2$ is the distance from the other focus. We can now define an ellipse. For each point on the ellipse, $(d_1 + d_2)$ is a constant. Let's call it $2a$.

$$d_1 + d_2 = 2a, \text{ where } d_1 = \sqrt{(x-c)^2 + y^2} \text{ and } d_2 = \sqrt{(x+c)^2 + y^2}.$$

$$d_1 = 2a - d_2$$
$$\sqrt{(x-c)^2 + y^2} = 2a - \sqrt{(x+c)^2 + y^2}$$
$$(x-c)^2 + y^2 = 4a^2 - 4a\sqrt{(x+c)^2 + y^2} + (x+c)^2 + y^2$$
$$x^2 - 2cx + c^2 + y^2 = 4a^2 - 4a\sqrt{(x+c)^2 + y^2} + x^2 + 2cx + c^2 + y^2$$
$$-4cx - 4a^2 = -4a\sqrt{(x+c)^2 + y^2}$$
$$cx + a^2 = a\sqrt{(x+c)^2 + y^2}$$
$$(cx + a^2)^2 = a^2\left((x+c)^2 + y^2\right)$$
$$c^2x^2 + 2ca^2x + a^4 = a^2x^2 + 2ca^2x + a^2c^2 + a^2y^2$$
$$a^2\left(a^2 - c^2\right) = x^2\left(a^2 - c^2\right) + a^2y^2$$

By definition, $d_1 + d_2 = 2a$. From the above figure, we see that $d_1 + d_2 > 2c$. Therefore $a > c$. This makes $a^2 - c^2 > 0$. Hence we can divide by $\left(a^2 - c^2\right)$.

$$a^2 = x^2 + \frac{a^2y^2}{\left(a^2 - c^2\right)} \quad \text{or} \quad 1 = \frac{x^2}{a^2} + \frac{y^2}{\left(a^2 - c^2\right)}.$$

If we let $b^2 = \left(a^2 - c^2\right)$, we get the standard equation for a horizontal ellipse.

$$\frac{x^2}{a^2} + \frac{y^2}{b^2} = 1$$

Equation E.1

Note that $a > b$. If $x = 0$, then $y = \pm b$. Also, if $y = 0$, then $x = \pm a$. See the following figure.

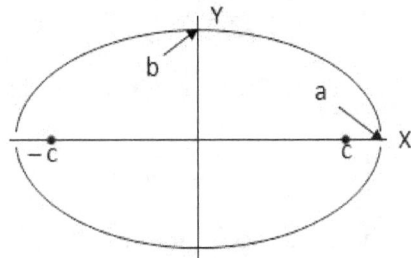

As can be seen, if we are given the plot, it's easy to measure a and b, and then compute c.

Let's shift the ellipse so that the left focus is at the origin. This is done by substituting $x = x - c$ into equation E.1.

$$\frac{(x-c)^2}{a^2} + \frac{y^2}{b^2} = 1$$

Equation E.2

See the following figure.

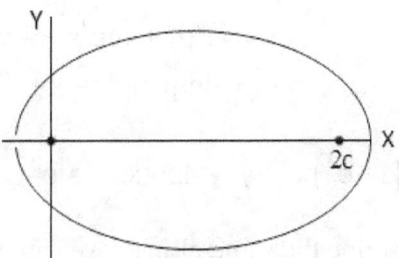

With one focus at the origin, we can derive the polar equation for a horizontal ellipse. As with the parabola, we draw what is called a directrix. This is a line that is perpendicular to the horizontal axis. This one will be located at $x = -2p$. (We will discuss the value of p in a little while.) The following figure shows the ellipse, the directrix, and two distances that will be used to define the ellipse.

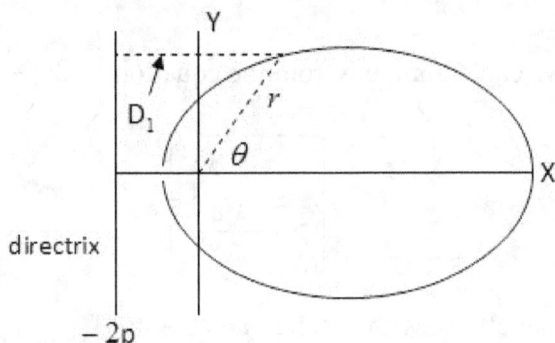

$D_1$ is the distance from the directrix to any point on the ellipse. $r$ is the distance from the left focus to that same point. In rectangular coordinates, all points on the ellipse satisfy $d_1 + d_2 = 2a$. In polar coordinates, all points on the ellipse satisfy the following equation.

$$\frac{r}{D_1} = (\text{a constant which we will call e}).$$

From the above figure, $D_1 = 2p + x$. Therefore,

$$\frac{r}{D_1} = \frac{r}{2p + x} = e$$

$$r = e(2p + x) = 2ep + er\cos\theta$$

$$r = \frac{2ep}{1 - e\cos\theta} \qquad\qquad \text{Equation E.3}$$

Equations E.2 and E.3 represent the same horizontal ellipse. To show this, we will put each equation into what is called the standard form.

Rewriting equation E.2, $b^2(x-c)^2 + a^2y^2 - a^2b^2 = 0$. This yields the standard form.

$$\left(\frac{b}{a}\right)^2 x^2 + y^2 - 2c\left(\frac{b}{a}\right)^2 x + c^2\left(\frac{b}{a}\right)^2 - b^2 = 0 \qquad\qquad \text{Equation E.4}$$

Rewriting equation E.3, $r(1 - e\cos\theta) = 2ep$. Therefore, $r = 2ep + ex$.

$$r^2 = 4e^2p^2 + 4pe^2x + e^2x^2$$

$$x^2 + y^2 = 4e^2p^2 + 4pe^2x + e^2x^2$$

This yields the standard form.

$$(1 - e^2)x^2 + y^2 - 4pe^2x - 4c^2p^2 = 0 \qquad\qquad \text{Equation E.5}$$

Since equations E.4 and E.5 describe the same ellipse, we can equate the coefficients.

- From the $x^2$ coefficient, we can compute e from the equation $\left(\left(\frac{b}{a}\right)^2 = 1 - e^2\right)$. This yields $\boxed{e = \frac{c}{a}}$.

  The constant e is called the eccentricity. Since $a > c$, $e < 1$ for an ellipse.

- From the x coefficient, we can compute p from the equation $\left(-2c\left(\frac{b}{a}\right)^2 = -4pe^2\right)$. This yields

$$\boxed{p = \frac{c\left(\frac{b}{a}\right)^2}{2e^2}}$$

Suppose we want to make an ellipse with $a = 2.22$ and $b = 1.33$.

- From $c^2 = a^2 - b^2$, we compute $c = 1.8$. The focus is $(a - c) = 0.44$ from the vertex.
- The eccentricity is $e = \frac{c}{a} = 0.8$.

- For the location of the directrix, $p = \dfrac{c\left(\dfrac{b}{a}\right)^2}{2e^2} = 0.5$ .

We use these values to construct the ellipse in the following figure using both rectangular and polar coordinates.

| Horizontal Ellipse with the Origin at the Left Focus | |
|---|---|
| **Rectangular Coordinates Equation E.2** $$\frac{(x-c)^2}{a^2} + \frac{y^2}{b^2} = \frac{(x-1.8)^2}{(2.22)^2} + \frac{y^2}{(1.33)^2} = 1$$ $$\left(\frac{b}{a}\right)^2 x^2 + y^2 - 2c\left(\frac{b}{a}\right)^2 x + c^2\left(\frac{b}{a}\right)^2 - b^2 = 0$$ **Polar Coordinates Equation E.3** $$r = \frac{2pe}{1 - e\cos\theta} = \frac{2*0.5*0.8}{1 - 0.8\cos\theta}$$ $$\left(1 - e^2\right)x^2 + y^2 - 4pe^2 x - 4p^2 e^2 = 0$$ **Plugging in Numbers** $$0.36x^2 + y^2 - 1.28x - 0.64 = 0$$ | 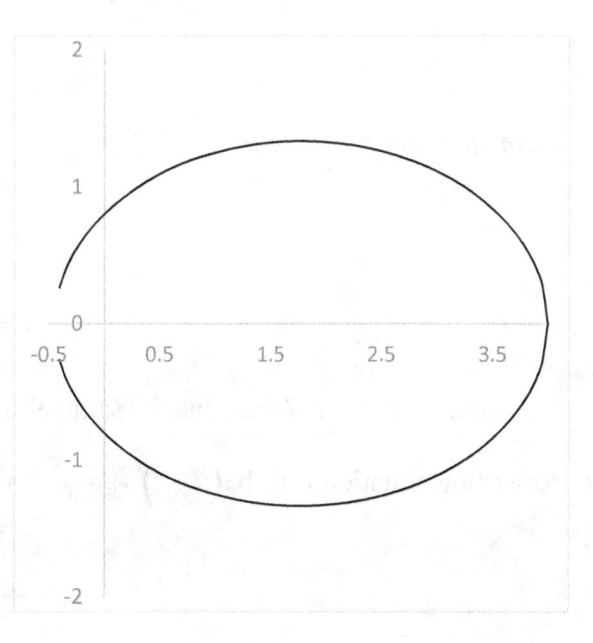 |

Still dealing with the horizontal ellipse, substituting $x = x + c$ into equation E.1 will shift the origin to the right focus.

$$\frac{(x+c)^2}{a^2} + \frac{y^2}{b^2} = 1 \qquad\qquad \text{Equation E.6}$$

The following is the resulting plot.

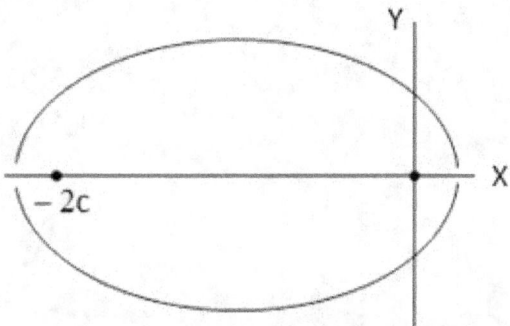

153

Now let's draw a directrix at $x = 2p$. The following is the resulting plot.

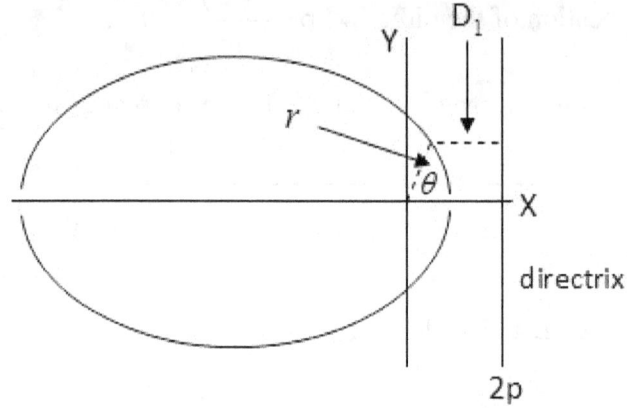

Here $D_1 = 2p - x$. Therefore,

$$\frac{r}{D_1} = \frac{r}{2p - x} = e$$

$$r = \frac{2ep}{1 + e\cos\theta} \qquad\qquad \text{Equation E.7}$$

Equations E.6 and E.7 describe the same ellipse. Let's put them into standard form.

Rewriting equation E.6: $b^2(x+c)^2 + a^2y^2 - a^2b^2 = 0$. This yields the standard form.

$$\left(\frac{b}{a}\right)^2 x^2 + y^2 + 2c\left(\frac{b}{a}\right)^2 x + c^2\left(\frac{b}{a}\right)^2 - b^2 = 0 \qquad\qquad \text{Equation E.8}$$

Rewriting equation E.7: $r(1 + e\cos\theta) = 2ep$. Therefore: $r = 2ep - ex$. This yields

$$(1 - e^2)x^2 + y^2 + 4pe^2x - 4e^2p^2 = 0 \qquad\qquad \text{Equation E.9}$$

Equations E.8 and E.9 will yield the same parameter values as did equations E.4 and E.5. The only difference is that the right focus will be at the origin. This can be seen in the following table.

| Horizontal Ellipse with the Origin at the Right Focus | |
|---|---|
| <u>Rectangular Coordinates, Equation E.6</u><br><br>$$\frac{(x+c)^2}{a^2}+\frac{y^2}{b^2}=\frac{(x+1.8)^2}{(2.22)^2}+\frac{y^2}{(1.33)^2}=1$$<br><br>$$\left(\frac{b}{a}\right)^2 x^2+y^2+2c\left(\frac{b}{a}\right)^2 x+c^2\left(\frac{b}{a}\right)^2-b^2=0$$<br><br><u>Polar Coordinates, Equation E.7</u><br><br>$$r=\frac{2pe}{1+e\cos\theta}=\frac{2*0.5*0.8}{1+0.8\cos\theta}$$<br><br>$$(1-e^2)x^2+y^2+4pe^2x-4p^2e^2=0$$<br><br><u>Plugging in Numbers</u><br>$$0.36x^2+y^2+1.28x-0.64=0$$ | 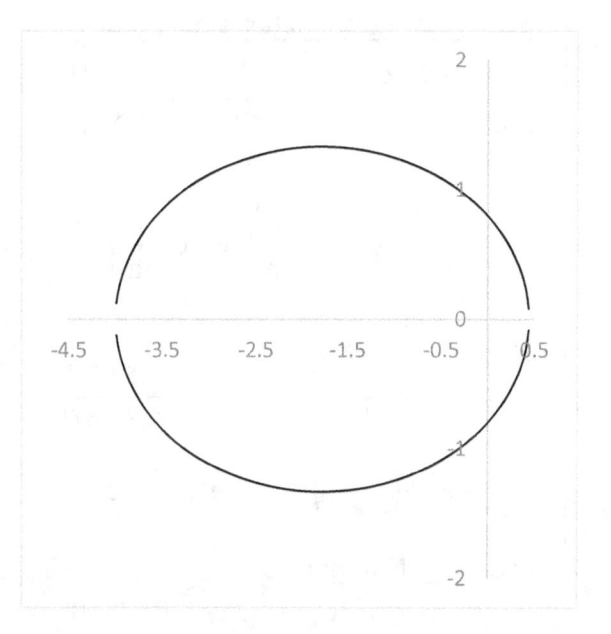 |

## Vertical Ellipse

With the parabola, we showed we could turn the horizontal equations into vertical equations by doing the following.

- In rectangular coordinates, reverse the roles of x and y.
- In polar coordinates, replace cosθ with sinθ.

The same is true for the ellipse as shown in the following three tables.

| Vertical Ellipse | |
|---|---|
| <u>Rectangular Coordinates</u><br><br>$$\frac{x^2}{b^2}+\frac{y^2}{a^2}=\frac{x^2}{(1.33)^2}+\frac{y^2}{(2.22)^2}=1$$<br><br>$$x^2+\left(\frac{b}{a}\right)^2 y^2-b^2=0$$<br><br><u>Plugging in Numbers</u><br>$$x^2+0.36y^2-1.8=0$$ | |

155

## Vertical Ellipse with the Origin at the Bottom Focus

**Rectangular Coordinates**

$$\frac{x^2}{b^2}+\frac{(y-c)^2}{a^2} = \frac{x^2}{(1.33)^2}+\frac{(y-1.8)^2}{(2.22)^2} = 1$$

$$x^2+\left(\frac{b}{a}\right)^2 y^2 - 2c\left(\frac{b}{a}\right)^2 y + c^2\left(\frac{b}{a}\right)^2 - b^2 = 0$$

**Polar Coordinates**

$$r = \frac{2pe}{1-e\sin\theta} = \frac{2*0.5*0.8}{1-0.8\sin\theta}$$

$$x^2 + \left(1 - e^2\right)y^2 - 4pe^2 y - 4p^2 e^2 = 0$$

**Plugging in Numbers**

$$x^2 + 0.36y^2 - 1.28y - 0.64 = 0$$

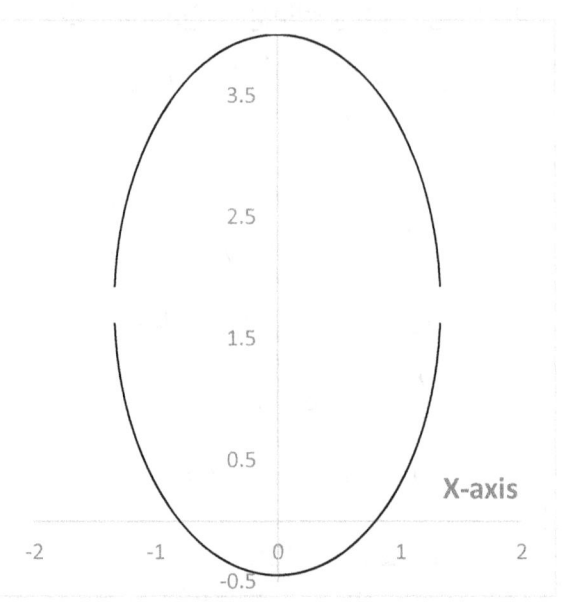

## Vertical Ellipse with the Origin at the Top Focus

**Rectangular Coordinates**

$$\frac{x^2}{b^2}+\frac{(y+c)^2}{a^2} = \frac{x^2}{(1.33)^2}+\frac{(y+1.8)^2}{(2.22)^2} = 1$$

$$x^2+\left(\frac{b}{a}\right)^2 y^2 + 2c\left(\frac{b}{a}\right)^2 y + c^2\left(\frac{b}{a}\right)^2 - b^2 = 0$$

**Polar Coordinates**

$$r = \frac{2pe}{1+e\sin\theta} = \frac{2*0.5*0.8}{1+0.8\sin\theta}$$

$$x^2 + \left(1 - e^2\right)y^2 + 4pe^2 y - 4p^2 e^2 = 0$$

**Plugging in Numbers**

$$x^2 + 0.36y^2 + 1.28y - 0.64 = 0$$

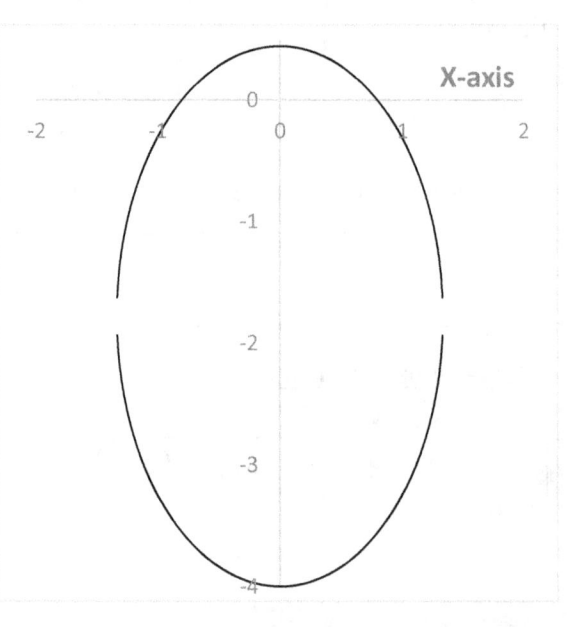

# Hyperbola

A hyperbola has two foci. As shown in the following figure, they are at $\pm c$.

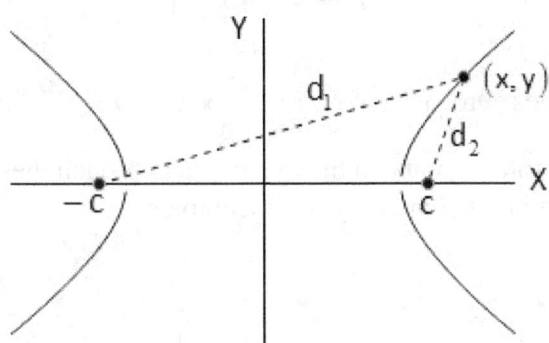

Also shown are distances $d_1$ and $d_2$. We now define a hyperbola. For each point on the hyperbola,

$$|d_1 - d_2| \text{ is a positive constant which we will call } 2a.$$

This means that $(d_1 - d_2 = 2a)$ or $(d_2 - d_1 = 2a)$. Let's pick $(d_1 - d_2 = 2a)$. From the figure we see

$$d_1 = \sqrt{(x+c)^2 + y^2} \text{ and } d_2 = \sqrt{(x-c)^2 + y^2}.$$

We start the derivation with

$$d_1 = 2a + d_2$$
$$d_1^2 = 4a^2 + 4ad_2 + d_2^2$$
$$d_1^2 - d_2^2 - 4a^2 = 4ad_2$$
$$d_1^2 - d_2^2 - 4a^2 = 4a\sqrt{(x-c)^2 + y^2}$$
$$4cx - 4a^2 = 4a\sqrt{(x-c)^2 + y^2}$$
$$c^2x^2 - 2cxa^2 + a^4 = a^2\left(x^2 - 2cx + c^2 + y^2\right)$$
$$x^2\left(c^2 - a^2\right) - a^2y^2 = a^2\left(c^2 - a^2\right)$$

Let's let $b^2 = \left(c^2 - a^2\right)$. Then, $\qquad x^2b^2 - a^2y^2 = a^2b^2$.

If $y = 0$, then $x = \pm a$. Therefore, $c > a$ and $\left(c^2 - a^2\right) = b^2 > 0$. Hence we can divide by $\left(a^2b^2\right)$ to yield the familiar equation for a horizontal hyperbola centered at the origin.

$$\frac{x^2}{a^2} - \frac{y^2}{b^2} = 1 \qquad\qquad \text{Equation H.1}$$

Note: Equation H.1 can also be derived from $(d_2 - d_1 = 2a)$.

Let's examine equation H.1. If $x = 0$, then $y = \pm jb$. This means that this hyperbola does not cross the y-axis. Solving equation H.1 for $y^2$,

$$y^2 = \left(\frac{b}{a}x\right)^2\left(1 - \left(\frac{a}{x}\right)^2\right)$$

For large x, $\left(\frac{a}{x}\right)^2$ approaches zero. This leaves $y = \frac{b}{a}x$ and $y = -\frac{b}{a}x$. These two equations are straight lines that are called asymptotes. Points on the hyperbola approach these lines for large x. The following figure shows how a hyperbola is shaped by its asymptotes.

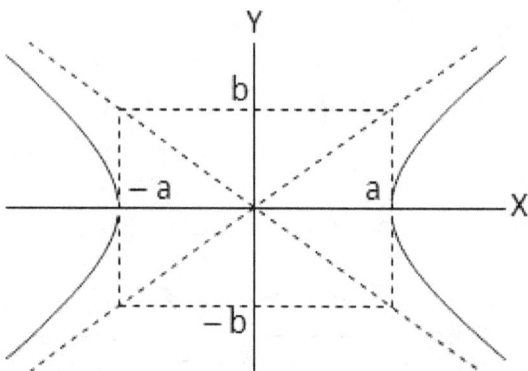

What if you are given the plot without the values of **a** and **b**? The value of **a** can be quite accurately obtained from the plot. But asymptotes are generally not accurate enough to determine the value of **b**. Let's solve equation H.1 for **b**.

$$b - \frac{ay}{\sqrt{x^2 - a^2}}$$

Equation H.2

From the plot, you can pick off the value of y for any value of x. You can use these to compute b.

The following table shows a horizontal hyperbola that is centered at the origin. The value of **a** is easily determined. The figure also shows a point at x = 4.5. This point was used to compute the value of **b** from equation H.2.

Equation H.1

$$\frac{x^2}{a^2} - \frac{y^2}{b^2} = \frac{x^2}{(2.723)^2} - \frac{y^2}{(1.81)^2} = 1$$

Rewriting

$$-\left(\frac{b}{a}\right)^2 x^2 + y^2 + b^2 = 0$$

Plugging in Numbers

$$-0.44x^2 + y^2 + 3.28 = 0$$

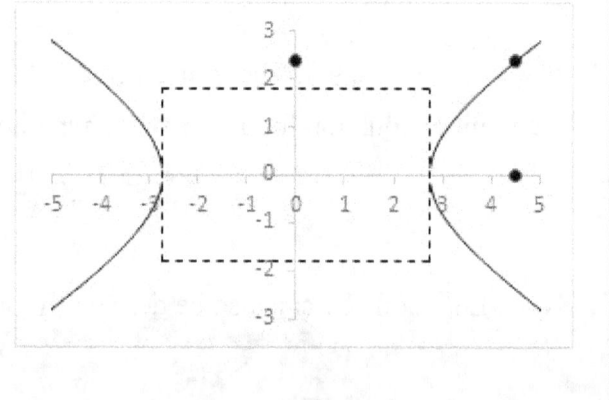

Now let's move the right focus to the origin. This is done by substituting $x = x + c$ into equation H.1.

$$\frac{(x+c)^2}{a^2} - \frac{y^2}{b^2} = 1 \qquad\qquad \text{Equation H.3}$$

The following figure shows this.

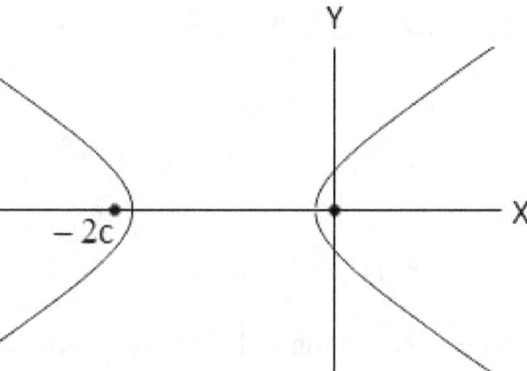

With one focus at the origin, we will derive the polar equation that is equivalent to equation H.3. We will do this using the following figure.

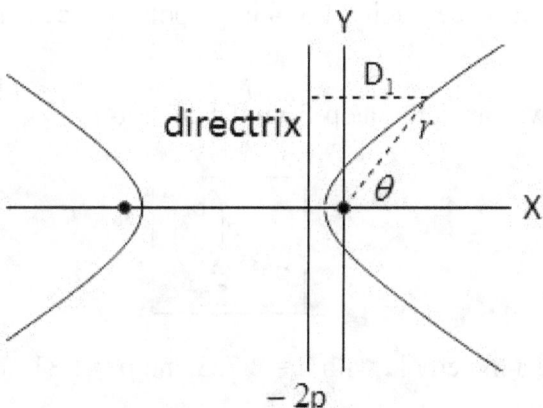

The figure shows a directrix that is perpendicular to the x-axis at $x = -2p$. The figure also shows two distances, $D_1$ and $r$. In rectangular coordinates, the hyperbola was defined using the equation $|d_1 - d_2| = 2a$. In polar coordinates, it is defined using the equation

$$\frac{r}{D_1} = \left(\text{a constant which we will call e}\right).$$

From the figure, we see that $D_1 = 2p + x$. Therefore, $\dfrac{r}{D_1} = \dfrac{r}{2p + x} = e$. This can be manipulated into

$$r = \frac{2pe}{1 - e\cos\theta} \qquad\qquad \text{Equation H.4}$$

To show how equations H.3 and H.4 represent the same hyperbola, we will put them into a standard equation form.

Rewriting equation H.3, $b^2(x+c)^2 - a^2y^2 - a^2b^2 = 0$. This yields

$$-\left(\frac{b}{a}\right)^2 x^2 + y^2 - 2c\left(\frac{b}{a}\right)^2 x - c^2\left(\frac{b}{a}\right)^2 + b^2 = 0 \qquad \text{Equation H.5}$$

Rewriting equation H.4, $r(1 - e\cos\theta) = 2ep$. Therefore, $r = 2ep + ex$.

$$r^2 = 4e^2p^2 + 4pe^2x + e^2x^2$$
$$x^2 + y^2 = 4e^2p^2 + 4pe^2x + e^2x^2$$

This yields the standard form

$$(1 - e^2)x^2 + y^2 - 4pe^2x - 4e^2p^2 = 0 \qquad \text{Equation H.6}$$

Since equations H.5 and H.6 describe the same ellipse, we can equate the coefficients.

- From the $x^2$ coefficient, we can compute e from the equation $\left(-\left(\frac{b}{a}\right)^2 = 1 - e^2\right)$. This yields $\boxed{e = \frac{c}{a}}$.

  The constant e is called the eccentricity. For a hyperbola, $a < c$, which means that $e > 1$.

- From the x coefficient, we can compute p from the equation $\left(-2c\left(\frac{b}{a}\right)^2 = -4pe^2\right)$. This yields

$$\boxed{p = \frac{c\left(\frac{b}{a}\right)^2}{2e^2}}$$

Suppose we want to make a hyperbola with $a = 2.723$ and $b = 1.81$.

- From $c^2 = a^2 + b^2$, we compute $c = 3.27$. The focus, $(c - a)$, is 0.55 from the vertex.

- The eccentricity is $e = \frac{c}{a} = 1.2$.

- For the location of the directrix, $p = \frac{c(b/a)^2}{2e^2} = 0.5$.

We use these values to construct the hyperbola in the following figure using both rectangular and polar coordinates.

## Horizontal Hyperbola with the Origin at the Right Focus

Rectangular Coordinates, Equation H.3

$$\frac{(x+c)^2}{a^2} - \frac{y^2}{b^2} = \frac{(x+3.273)^2}{(2.723)^2} - \frac{y^2}{(1.81)^2} = 1$$

$$-\left(\frac{b}{a}\right)^2 x^2 + y^2 - 2c\left(\frac{b}{a}\right)^2 x - c^2\left(\frac{b}{a}\right)^2 + b^2 = 0$$

Polar Coordinates, Equation H.4

$$r = \frac{2pe}{1 - e\cos\theta} = \frac{2*0.5*1.2}{1 - 1.2\cos\theta}$$

$$\left(1 - e^2\right)x^2 + y^2 - 4pe^2x - 4p^2e^2 = 0$$

Plugging in Numbers

$$-0.44x^2 + y^2 - 2.88x - 1.44 = 0$$

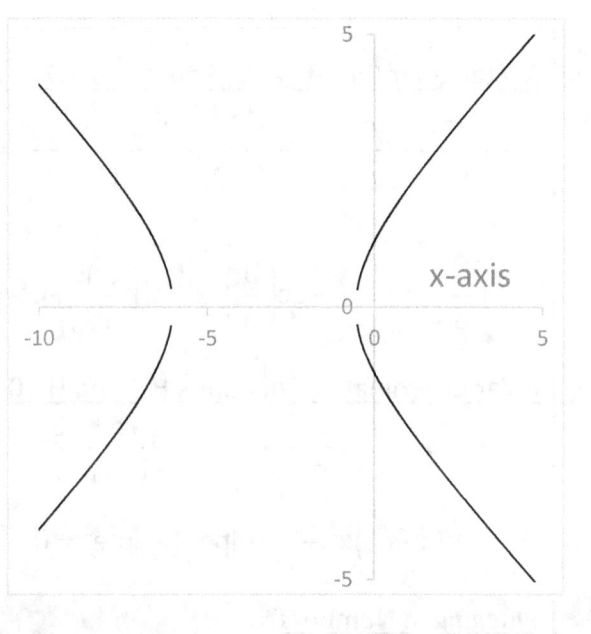

Let's shift the left focus to the origin. This is done by substituting x = ( x – c ) in equation H.1.

- In rectangular coordinates,     $$\frac{(x-c)^2}{a^2} - \frac{y^2}{b^2} = 1$$     Equation H.7

In equation form,     $$-\left(\frac{b}{a}\right)^2 x^2 + y^2 + 2c\left(\frac{b}{a}\right)^2 x - c^2\left(\frac{b}{a}\right)^2 + b^2 = 0$$     Equation H.8

- In polar coordinates, since $D_1 = 2p - r\cos\theta$,  $$r = \frac{2pe}{1 + e\cos\theta}$$     Equation H.9

In equation form,     $$\left(1 - e^2\right)x^2 + y^2 + 4pe^2x - 4e^2p^2 = 0$$     Equation H.10

Equations H.8 and H.10 yield the same results for eccentricity (e) and directrix location (p) that were previously derived.

The following table summarizes the data for a horizontal hyperbola with the left focus at the origin.

| Horizontal Hyperbola with the Origin at the Left Focus | |
|---|---|
| Rectangular Coordinates, Equations H.7 and H.8<br><br>$$\frac{(x-c)^2}{a^2} - \frac{y^2}{b^2} = \frac{(x-3.273)^2}{(2.723)^2} - \frac{y^2}{(1.81)^2} = 1$$<br><br>$$-\left(\frac{b}{a}\right)^2 x^2 + y^2 + 2c\left(\frac{b}{a}\right)^2 x - c^2\left(\frac{b}{a}\right)^2 + b^2 = 0$$<br><br>Polar Coordinates, Equations H.9 and H.10<br><br>$$r = \frac{2pe}{1+e\cos\theta} = \frac{2*0.5*1.2}{1+1.2\cos\theta}$$<br><br>$$\left(1-e^2\right)x^2 + y^2 + 4pe^2 x - 4p^2e^2 = 0$$<br><br>Plugging in Numbers<br>$$-0.44x^2 + y^2 + 2.88x - 1.44 = 0$$ | 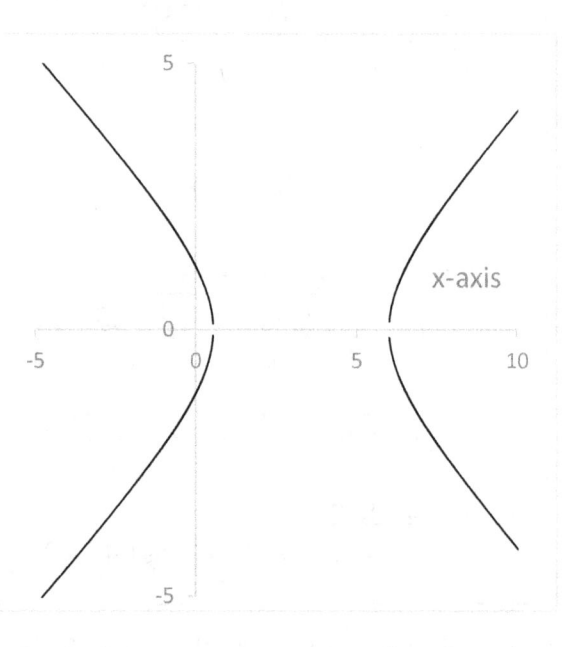 |

## Vertical Hyperbola

We can turn the horizontal equations into vertical equations by doing the following.

- In rectangular coordinates, reverse the roles of x and y.
- In polar equations, replace cosθ with sinθ.

This is shown in the following three tables.

| Vertical Hyperbola Centered at the Origin | |
|---|---|
| Rectangular Coordinates<br><br>$$\frac{y^2}{a^2} - \frac{x^2}{b^2} = \frac{y^2}{(2.723)^2} - \frac{x^2}{(1.81)^2} = 1$$<br><br>$$x^2 - \left(\frac{b}{a}\right)^2 y^2 + b^2 = 0$$<br><br>Plugging in Numbers<br>$$x^2 - 0.44y^2 + 3.28 = 0$$ | (graph of vertical hyperbola with x-axis label) |

## Vertical Hyperbola with the Origin at the Top Focus

### Rectangular Coordinates

$$\frac{(y+c)^2}{a^2} - \frac{x^2}{b^2} = \frac{(y+3.273)^2}{(2.723)^2} - \frac{x^2}{(1.81)^2} = 1$$

$$x^2 - \left(\frac{b}{a}\right)^2 y^2 - 2c\left(\frac{b}{a}\right)^2 y - c^2\left(\frac{b}{a}\right)^2 + b^2 = 0$$

### Polar Coordinates

$$r = \frac{2pe}{1 - e\sin\theta} = \frac{2*0.5*1.2}{1 - 1.2\sin\theta}$$

$$x^2 + \left(1 - e^2\right)y^2 - 4pe^2 y - 4p^2 e^2 = 0$$

### Plugging in Numbers

$$x^2 - 0.44y^2 - 2.88y - 1.44 = 0$$

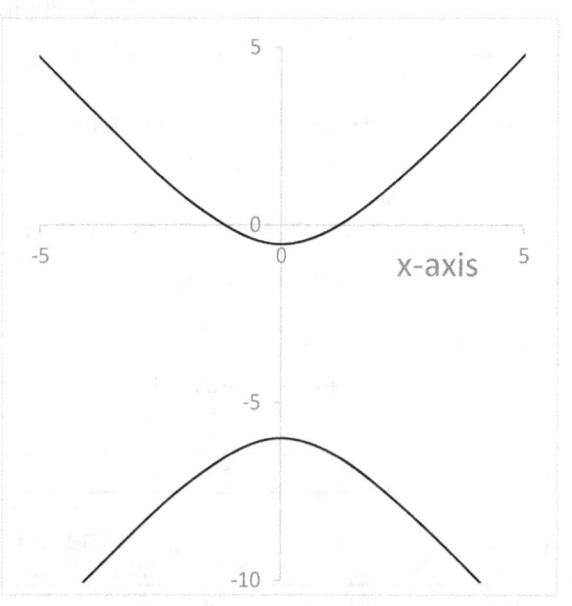

## Vertical Hyperbola with the Origin at the Bottom Focus

### Rectangular Coordinates

$$\frac{(y-c)^2}{a^2} - \frac{x^2}{b^2} = \frac{(y-3.273)^2}{(2.723)^2} - \frac{x^2}{(1.81)^2} = 1$$

$$x^2 - \left(\frac{b}{a}\right)^2 y^2 + 2c\left(\frac{b}{a}\right)^2 y - c^2\left(\frac{b}{a}\right)^2 + b^2 = 0$$

### Polar Coordinates

$$r = \frac{2pe}{1 + e\sin\theta} = \frac{2*0.5*1.2}{1 + 1.2\sin\theta}$$

$$x^2 + \left(1 - e^2\right)y^2 + 4pe^2 y - 4p^2 e^2 = 0$$

### Plugging in Numbers

$$x^2 - 0.44y^2 + 2.88y - 1.44 = 0$$

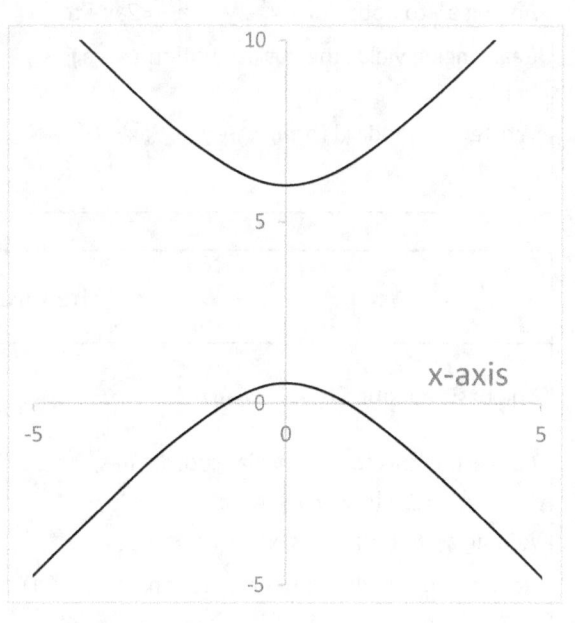

# Circle

The equations of a circle are summarized in the following three tables.

| The Circle in Rectangular Coordinates | |
| --- | --- |
| The basic equation of a circle centered at the origin is $$x^2 + y^2 = a^2$$ When centered at $x = a$ and $y = b$, the equation is $$(x-a)^2 + (y-b)^2 = a^2$$ Written in standard form, $$x^2 + y^2 - 2ax - 2by + b^2 = 0$$ | 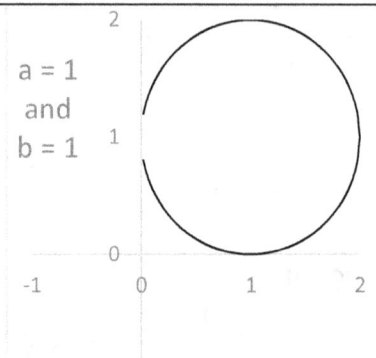 |

| The Circle in Polar Cosine Form | |
| --- | --- |
| The basic equation is $r = 2a\cos\theta$. Converting this to rectangular coordinates, $$r^2 = 2a\cos\theta * r \quad \text{or} \quad x^2 + y^2 = 2ax.$$ Adding $a^2$ to both sides $x^2 + y^2 + a^2 = 2ax + a^2$. Rearranging yields the basic equation $(x-a)^2 + y^2 = a^2$ Written in standard form, $x^2 + y^2 - 2ax = 0$. | 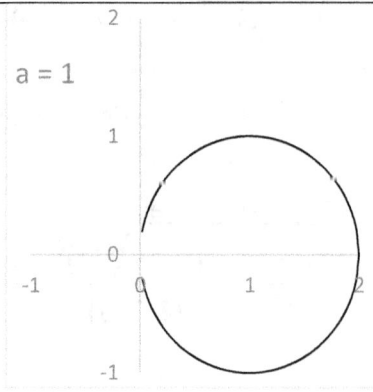 |

| The Circle in Polar Sine Form | |
| --- | --- |
| The basic equation is $r = 2a\sin\theta$. Converting this to rectangular coordinates, $$r^2 = 2a\sin\theta * r \quad \text{or} \quad x^2 + y^2 = 2ay.$$ Adding $a^2$ to both sides $x^2 + y^2 + a^2 = 2ay + a^2$. Rearranging yields the basic equation $x^2 + (y-a)^2 = a^2$ Written in standard form, $x^2 + y^2 - 2ay = 0$. | 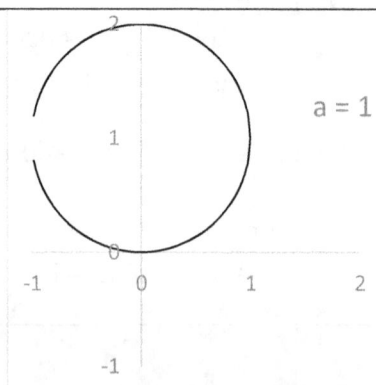 |

# Translating Conic Equations

In the previous sections, when we were comparing figures in polar coordinates with those in rectangular coordinates, we translated foci to the origin. In this section, we will illustrate general translation. The following table summarizes the possibilities.

Assume that h and k are positive numbers.

If we replace x by $(x-h)$, we will get the original figure shifted h units to the right.

If we replace x by $(x+h)$, we will get the original figure shifted h units to the left.

If we replace y by $(y-k)$, we will get the original figure shifted k units upward.

If we replace y by $(y+k)$, we will get the original figure shifted k units downward.

The following two tables illustrate this for a horizontal parabola.

### Table P.1: Horizontal Parabola, Centered

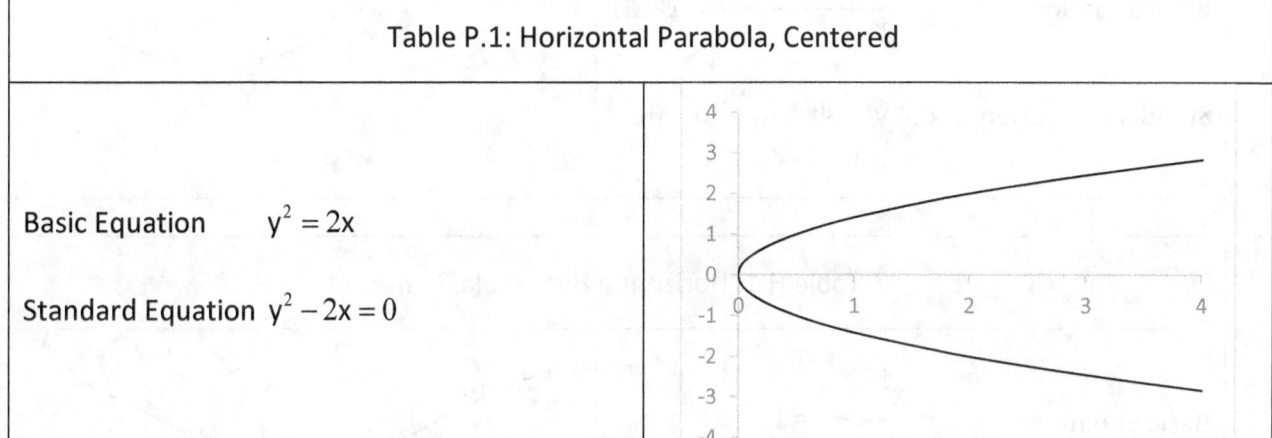

Basic Equation     $y^2 = 2x$

Standard Equation $y^2 - 2x = 0$

### Table P.2: Horizontal Parabola, Translated

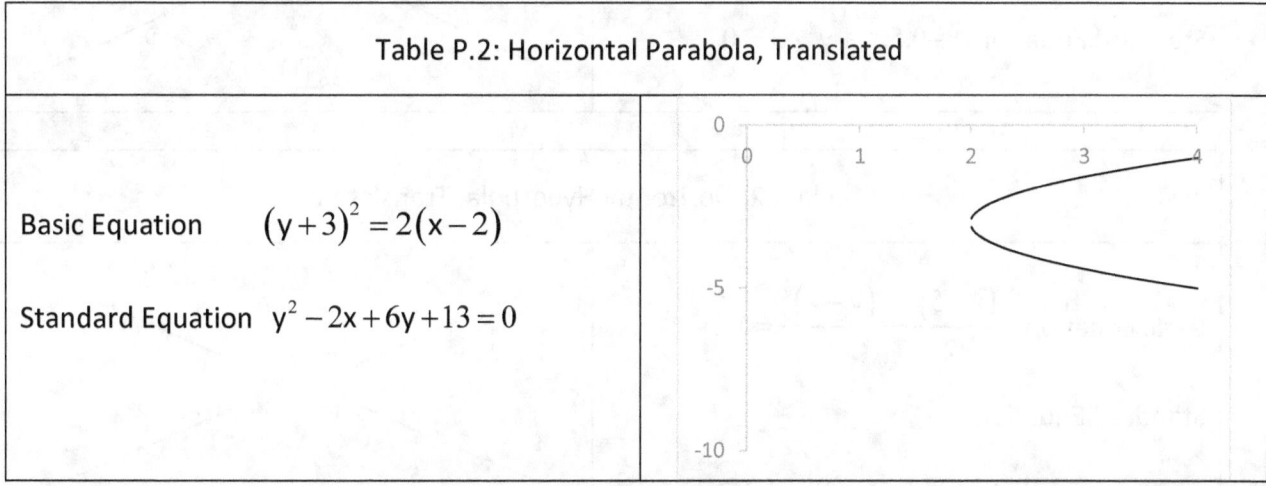

Basic Equation     $(y+3)^2 = 2(x-2)$

Standard Equation $y^2 - 2x + 6y + 13 = 0$

The following four tables illustrate translation for a horizontal ellipse and hyperbola.

| Table E.1: Horizontal Ellipse, Centered | |
| --- | --- |
| Basic Equation $\dfrac{x^2}{2}+\dfrac{y^2}{1}=1$ <br><br> Standard Equation $.5x^2+y^2-1=0$ | 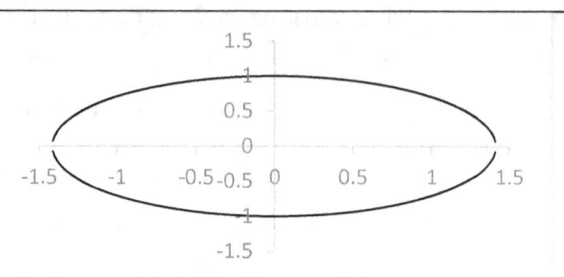 |

| Table E.2: Horizontal Ellipse, Translated | |
| --- | --- |
| Basic Equation $\dfrac{(x-4)^2}{2}+\dfrac{(y-3)^2}{1}=1$ <br><br> Standard Equation $.5x^2+y^2-4x-6y+16=0$ | 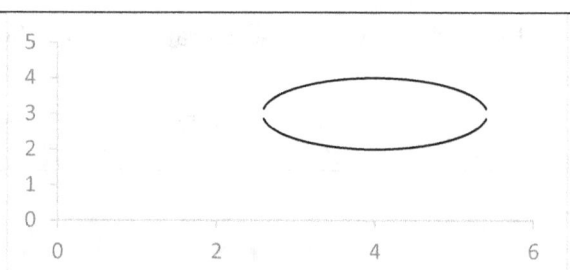 |

| Table H.1: Horizontal Hyperbola, Centered | |
| --- | --- |
| Basic Equation $\dfrac{x^2}{2}-\dfrac{y^2}{1}=1$ <br><br> Standard Equation $-0.5x^2+y^2+1=0$ | 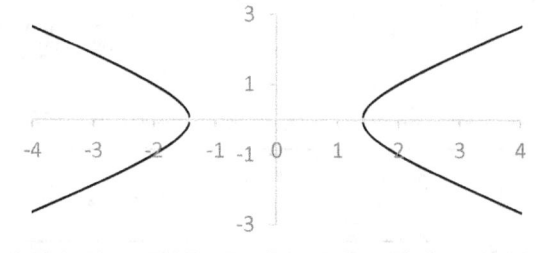 |

| Table H.2: Horizontal Hyperbola, Translated | |
| --- | --- |
| Basic Equation $\dfrac{(x-5)^2}{2}-\dfrac{(y-2)^2}{1}=1$ <br><br> Standard Equation <br><br> $-0.5x^2+y^2+5x-4y-7.5=0$ | 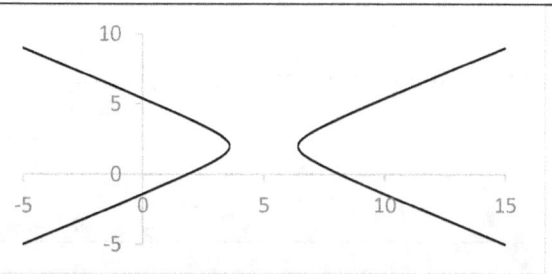 |

## Rotating Conic Equations

In this section, we will discuss how to rotate conic equations. From Table E.2, the standard equation for the translated ellipse is

$$.5x_p{}^2 + y_p{}^2 - 4x_p - 6y_p + 16 = 0 .$$

In this section, we will use the following notation for this type of equation:

$$A_p x_p^2 + C_p y_p^2 + D_p x_p + E_p y_p + F_p = 0$$

Suppose we are given an ellipse and its equation as shown in the following table.

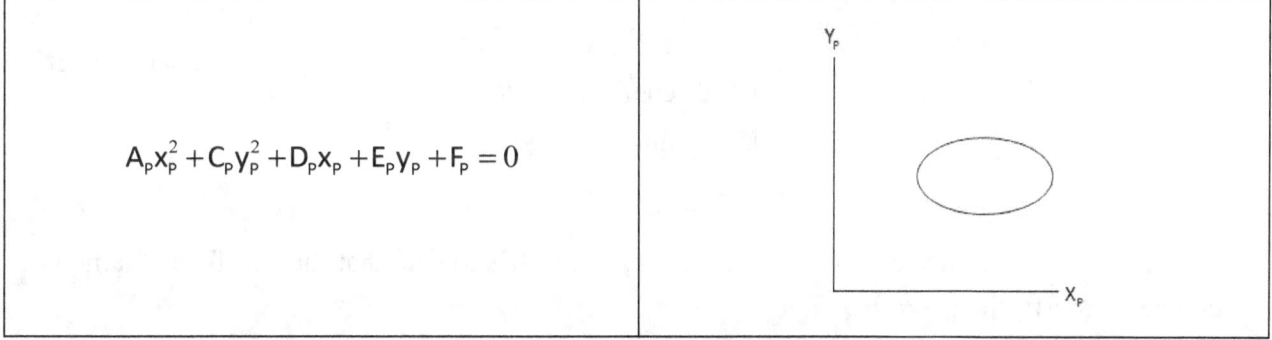

Let's define a new coordinate system, and rotate this figure as below.

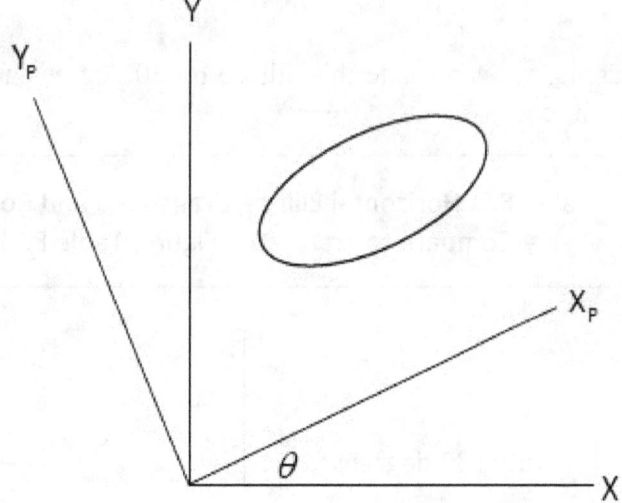

The following equations relating these two systems are derived in appendix E.

$$x_p = x\cos\theta + y\sin\theta$$
$$y_p = -x\sin\theta + y\cos\theta$$

Equation rotate

If we substitute these equations into the equations for the ellipse, we get:

$$A_p \left( x\cos\theta + y\sin\theta \right)^2 + C_p \left( -x\sin\theta + y\cos\theta \right)^2 + D_p \left( x\cos\theta + y\sin\theta \right) + E_p \left( -x\sin\theta + y\cos\theta \right) + F_p = 0$$

Performing the calculations and gathering the terms, we get the following equation in the new coordinate system:

$$Ax^2 + Bxy + Cy^2 + Dx + Ey + F = 0$$

where:

$$A = A_p \cos^2\theta + C_p \sin^2\theta$$
$$B = 2\sin\theta\cos\theta \left( A_p - C_p \right)$$
$$C = A_p \sin^2\theta + C_p \cos^2\theta$$
$$D = D_p \cos\theta - E_p \sin\theta$$
$$E = D_p \sin\theta + E_p \cos\theta$$
$$F = F_p$$

Equation coefficients

The first thing we notice is a new term, the $Bxy$ term. It's logical that the coefficient name is $B$. Of course, if $\theta = 0$, then $B = 0$.

We started this section with the equation of the ellipse from Table E.2:

$$.5x_p{}^2 + y_p{}^2 - 4x_p - 6y_p + 16 = 0$$

Using Equation coefficients, we can rotate this ellipse by 30 degrees about the origin. The result is shown in the following table:

| Table E.3: Horizontal Ellipse, Translated and Rotated |
|---|
| (Compare to Translated Figure, Table E.2) |

Basic Equation:
$$\frac{\left( x_p - 4 \right)^2}{2} + \frac{\left( y_p - 3 \right)^2}{1} = 1, \text{ rotated 30 degrees.}$$

Standard Equation:
$$.625x^2 - .433xy + .875y^2 - .464x - 7.196y + 16 = 0$$

The following table shows the parabola from Table P.2 after it is rotated −30 degrees.

| Table P.3: Horizontal Parabola, Translated and Rotated (Compare to Translated Figure, Table P.2) | |
| --- | --- |
| **Basic Equation** $$\left(y_p +3\right)^2 = 2\left(x_p -2\right),\text{ rotated (-30) degrees.}$$ **Standard Equation** $$.25x^2 +.866xy +.75y^2 +1.268x +6.196y +13 = 0$$ | 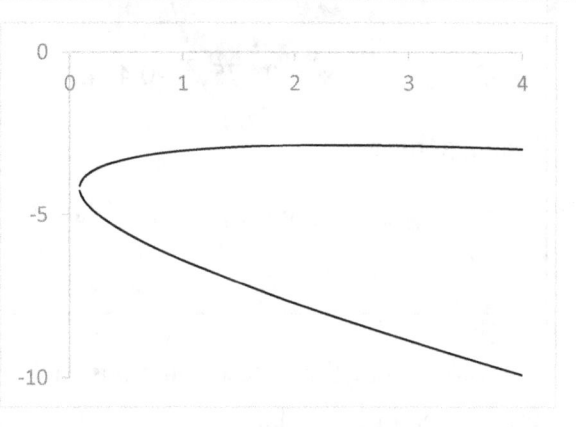 |

The following table shows the hyperbola from Table H.2 after it is rotated 30 degrees.

| Table H.3: Horizontal Hyperbola, Translated and Rotated (Compare to Translated Figure, Table H.2) | |
| --- | --- |
| **Basic Equation** $$\frac{\left(x_p -5\right)^2}{2} - \frac{\left(y_p -2\right)^2}{1} = 1,\text{ rotated 30 degrees.}$$ **Standard Equation** $$-0.125x^2 -1.3xy +0.625y^2 +6.33x -0.96y -7.5 = 0$$ | 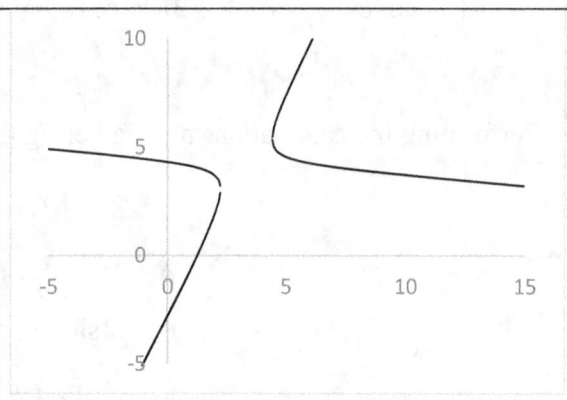 |

## Unrotating Conic Equations

Suppose we are given the equation of an ellipse in the x-y coordinate system of the previous section. The following table is from Table E.3.

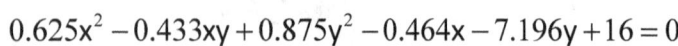

$$0.625x^2 - 0.433xy + 0.875y^2 - 0.464x - 7.196y + 16 = 0$$

Using general notation

$$Ax^2 + Bxy + Cy^2 + Dx + Ey + F = 0$$

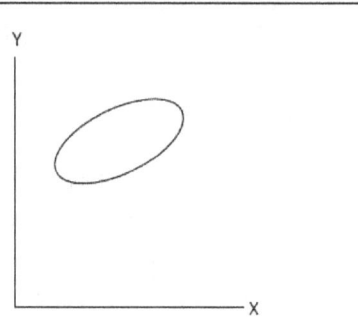

We want to transform these equations back into the $x_p - y_p$ coordinate system. The following equations are derived in appendix E.

$$x = x_p \cos\theta - y_p \sin\theta$$
$$y = x_p \sin\theta + y_p \cos\theta$$

If we substitute these equations into the equations for the ellipse, we get:

$$A\left(x_p \cos\theta - y_p \sin\theta\right)^2 + B\left(x_p \cos\theta - y_p \sin\theta\right)\left(x_p \sin\theta + y_p \cos\theta\right) + C\left(x_p \sin\theta + y_p \cos\theta\right)^2$$
$$+ D\left(x_p \cos\theta - y_p \sin\theta\right) + E\left(x_p \sin\theta + y_p \cos\theta\right) + F = 0$$

Performing the calculations and gathering the terms,

$$A_p x_p^2 + B_p x_p y_p + C_p y_p^2 + D_p x_p + E_p y_p + F_p = 0$$

where:

$$A_p = A\cos^2\theta + C\sin^2\theta + B\sin\theta\cos\theta$$
$$B_p = 2\sin\theta\cos\theta\left(C - A\right) + B\left(\cos^2\theta - \sin^2\theta\right)$$
$$C_p = A\sin^2\theta + C\cos^2\theta - B\sin\theta\cos\theta$$
$$D_p = D\cos\theta + E\sin\theta$$
$$E_p = -D\sin\theta + E\cos\theta$$
$$F_p = F$$

**Equation rotate back**

In the original $x_p$ - $y_p$ system, $B_p = 0$. We will now discuss how to force $B_p = 0$ when rotating back. From **Equation rotate back**, $B_p = 2\sin\theta\cos\theta\left(C - A\right) + B\left(\cos^2\theta - \sin^2\theta\right)$. We need to find the value of $\theta$ such that $B_p = 0$.

In the chapter on trigonometry, we derived two identities:

$$\sin(2\theta) = 2\sin\theta\cos\theta \quad \text{and} \quad \cos(2\theta) = \left(\cos^2\theta - \sin^2\theta\right)$$

Substituting these into $B_p$, we get $B_p = \sin(2\theta)(C - A) + B\cos(2\theta)$.

For $B_p = 0$, $\sin(2\theta)(A - C) = B\cos(2\theta)$. This means that

$$\tan(2\theta) = \frac{\sin(2\theta)}{\cos(2\theta)} = \frac{B}{(A-C)} \quad \text{or} \quad \theta = \frac{1}{2}\tan^{-1}\frac{B}{(A-C)}.$$

The following three tables show applications of this.

---

**Example 1: The Rotation Required to Eliminate the xy Term from Table P.3**

- Given the equation, $.25x^2 + .866xy + .75y^2 + 1.268x + 6.196y + 13 = 0$
- Compute the angle. $\theta = \frac{1}{2}\tan^{-1}\left(\frac{B}{A-C}\right) = \frac{1}{2}\tan^{-1}\left(\frac{0.866}{0.25 - 0.75}\right) = -30$ degrees
- Using this angle in **Equation rotate_back** yields $0x_p^2 + 0x_p y_p + y_p^2 - 2x_p + 6y_p + 13 = 0$
  This agrees with Table P.2.

---

**Example 2: The Rotation Required to Eliminate the xy Term from Table E.3**

- Given the equation, $.625x^2 - .433xy + .875y^2 - .464x - 7.196y + 16 = 0$
- Compute the angle. $\theta = \frac{1}{2}\tan^{-1}\left(\frac{B}{A-C}\right) = \frac{1}{2}\tan^{-1}\left(\frac{-0.433}{0.625 - 0.875}\right) = 30$ degrees
- Using this angle in **Equation rotate_back** yields $0.5x_p^2 + 0x_p y_p + y_p^2 - 4x_p - 6y_p + 16 = 0$
  This agrees with Table E.2.

---

**Example 3: The Rotation Required to Eliminate the xy Term from Table H.3**

- Given the equation $-0.125x^2 - 1.3xy + 0.625y^2 + 6.33x - 0.96y - 7.5 = 0$
- Compute the angle. $\theta = \frac{1}{2}\tan^{-1}\left(\frac{B}{A-C}\right) = \frac{1}{2}\tan^{-1}\left(\frac{-1.3}{-0.125 - 0.625}\right) = 30$ degrees
- Using this angle in **Equation rotate_back** yields $-0.5x_p^2 + 0x_p y_p + y_p^2 + 5x_p - 4y_p - 7.5 = 0$
  This agrees with Table H.2.

# Discriminants for Conic Equations

There are two cases for identifying the type of conic by looking at its equation.

**Case 1 - The Conic Is Not Rotated** $(B = 0)$: The following is its equation.

$$Ax^2 + Cy^2 + Dx + Ey + F = 0$$

The product AC is the discriminant.

> If $AC = 0$, the equation is a parabola because either $A = 0$ or $C = 0$.
> If $AC > 0$, the equation is an ellipse because A and C have the same sign.
> If $AC < 0$, the equation is a hyperbola because A and C have the opposite sign.

**Case 2 - The Conic Is Rotated** $B \neq 0$: The following is its equation.

$$Ax^2 + Bxy + Cy^2 + Dx + Ey + F = 0$$

Let's arrange this equation as if the value of x were known.

$$(C)y^2 + (Bx + E)y + (Ax^2 + Dx + F) = 0$$

Now let's use the quadratic formula to solve for y.

$$y = \frac{-(Bx + E) \pm \sqrt{(Bx + E)^2 - 4(C)(Ax^2 + Dx + F)}}{2(C)}$$

The terms can be rearranged.

$$y = \frac{-(Bx + E) \pm \sqrt{(B^2 - 4AC)x^2 + (2BE - 4CD)x + (E^2 - 4CF)}}{2(C)}$$

The product AC appears in the term $(B^2 - 4AC)$. When *rotating back*, we derived

$$A_p = A\cos^2\theta + C\sin^2\theta + B\sin\theta\cos\theta$$
$$B_p = \sin(2\theta)(C - A) + B\cos(2\theta)$$
$$C_p = A\sin^2\theta + C\cos^2\theta - B\sin\theta\cos\theta$$

Let's substitute these into the expression $(B_p^2 - 4A_pC_p)$.

$$B_p^2 - 4A_pC_p = (\sin(2\theta)(C - A) + B\cos(2\theta))^2$$
$$- 4(A\cos^2\theta + C\sin^2\theta + B\sin\theta\cos\theta)(A\sin^2\theta + C\cos^2\theta - B\sin\theta\cos\theta)$$

This boils down to $\left(B_p^2 - 4A_pC_p = B^2 - 4AC\right)$. We see that the value of $\left(B_p^2 - 4A_pC_p\right)$ isn't changed by rotation. When the conic is not rotated, $B_p = 0$. Therefore $\left(-4A_pC_p = B^2 - 4AC\right)$.

- For a parabola, $\left(B^2 - 4AC\right) = 0$ because $A_pC_p = 0$.
- For an ellipse, $\left(B^2 - 4AC\right) < 0$ because $A_pC_p > 0$.
- For a hyperbola, $\left(B^2 - 4AC\right) > 0$ because $A_pC_p < 0$.

The expression $\left(B^2 - 4AC\right)$ is the discriminant. The following three tables show that it works.

| Parabola | | |
|---|---|---|
| Type | Equation | $B^2 - 4AC$ |
| Centered (Table P.1) | $y^2 - 2x = 0$ | 0 |
| Translated (Table P.2) | $y^2 - 2x + 6y + 13 = 0$ | 0 |
| Translated-Rotated (Table P.3) | $.25x^2 + .866xy + .75y^2 + 1.268x + 6.196y + 13 = 0$ | 0 |

| Ellipse | | |
|---|---|---|
| Type | Equation | $B^2 - 4AC$ |
| Centered (Table E.1) | $.5x^2 + y^2 - 1 = 0$ | -2 |
| Translated (Table E.2) | $.5x^2 + y^2 - 4x - 6y + 16 = 0$ | -2 |
| Translated-Rotated (Table E.3) | $.625x^2 - .433xy + .875y^2 - .464x - 7.196y + 16 = 0$ | -2 |

| Hyperbola | | |
|---|---|---|
| Type | Equation | $B^2 - 4AC$ |
| Centered (Table H.1) | $-0.5x^2 + y^2 + 1 = 0$ | 2 |
| Translated (Table H.2) | $-0.5x^2 + y^2 + 5x - 4y - 7.5 = 0$ | 2 |
| Translated-Rotated (Table H.3) | $-0.125x^2 - 1.3xy + 0.625y^2 + 6.33x - 0.96y - 7.5 = 0$ | 2 |

## Interesting Cases

This chapter concludes with three cases that were encountered during this study.

---

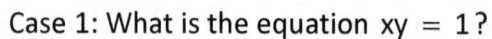

### Case 1: What is the equation $xy = 1$?

Answer

It's a hyperbola rotated 45°.

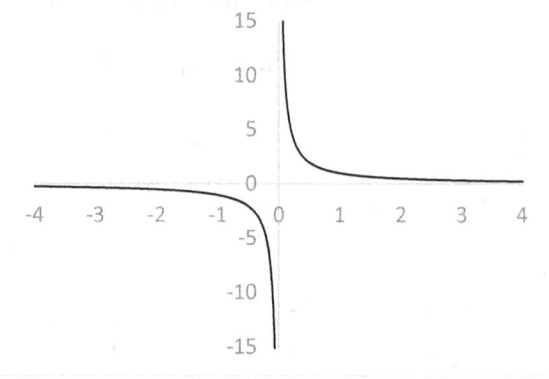

---

### Case 2: What is the equation $x^2 + 4xy + 4y^2 - 1 = 0$?

Answer

$$x^2 + 4xy + 4y^2 - 1 = \left(y + \frac{x}{2} + \frac{1}{2}\right)\left(y + \frac{x}{2} - \frac{1}{2}\right)$$

The equation factors into two parallel lines.

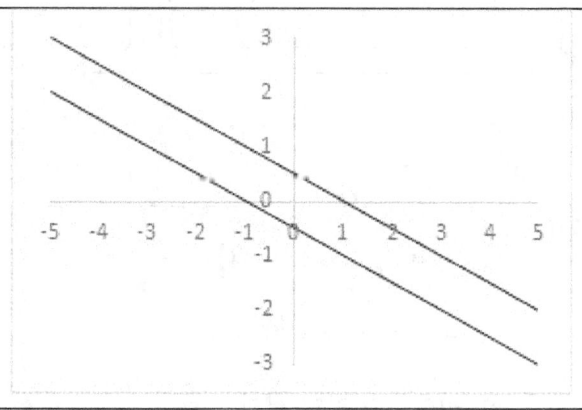

---

### Case 3: What is the equation $y^2 - \dfrac{x^2}{4} = 0$?

Answer

$$y^2 - \frac{x^2}{4} = \left(y + \frac{x}{2}\right)\left(y - \frac{x}{2}\right)$$

The equation factors into two intersecting lines.

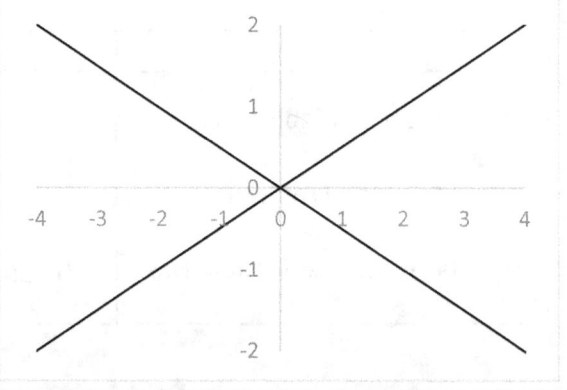

---

# The Conics – Programming in Chapter 7

Three programs were written to do this chapter. The first program is to make a plot of the standard equation of a conic:

$$Ax^2 + Bxy + Cy^2 + Dx + Ey + F = 0$$

Let's arrange this equation so that x can be the input.

$$Cy^2 + y(Bx+E) + (Ax^2 + Dx + F) = Cy^2 + y(Bx+E) + (X_{in}) = 0$$

If $C = 0$, $y = \dfrac{-X_{in}}{Bx+E}$. This is for a vertical parabola. In the program, we will call this y1.

If $C \neq 0$, $y1 = \dfrac{-(Bx+E) + \sqrt{(Bx+E)^2 - 4C*X_{in}}}{2C}$ and $y2 = \dfrac{-(Bx+E) - \sqrt{(Bx+E)^2 - 4C*X_{in}}}{2C}$.

The program will compute y1 and y2 for a range of x. The following is a listing of the program.

| Program *Sub general* to Plot $Ax^2 + Bxy + Cy^2 + Dx + Ey + F = 0$, Using x as the Input | |
|---|---|
| Sub general()<br><br>  A = Cells(5, 1): B = Cells(6, 1): C = Cells(7, 1)<br>  D = Cells(8, 1): E = Cells(9, 1): F = Cells(10, 1)<br><br>  Call equations(A, B, C, D, E, F)<br><br>End Sub  ' general<br><br>Sub equations(A, B, C, D, E, F)<br><br>  i = 1: Row = 0<br>  For x = -5 To 15 Step 0.02<br>    Xin = A * x ^ 2 + D * x + F<br>    If Abs(C) < 0.01 Then<br>      If Abs(B * x + E) < 0.01 Then<br>        Row = 1: GoTo bot<br>      End If<br>      y1 = -Xin / (B * x + E)<br>      y2 = 0<br>    Else<br>      rad = (B * x + E) ^ 2 - 4 * C * Xin<br>      If rad < 0 Then<br>        Row = 1: GoTo bot<br>      End If<br>      y1 = (-B * x - E + Sqr(rad)) / (2 * C)<br>      y2 = (-B * x - E - Sqr(rad)) / (2 * C)<br>    End If<br>    Cells(i + Row, 5) = x<br>    Cells(i + Row, 6) = y1<br>    Cells(i + Row, 7) = y2<br>    i = i + 1<br>bot:<br>  Next x<br><br>End Sub ' equations | • Read the coefficients.<br><br><br><br><br><br><br><br>• Row is for *plot penlifting* when transitioning between the segments of the hyperbola.<br><br><br><br>• If $(Bx + E) = 0$, the point x is not on the curve.<br><br><br><br>• y1 is for the vertical parabola.<br><br><br><br>• If $((Bx + E)^2 - 4CX_{in}) < 0$, the point x is not on the curve.<br><br><br><br>• y1 and y2 are for the ellipse, the hyperbola, the circle and the horizontal parabola. |

Program *Sub general* was used for the centered and translated figures. The following table is an example.

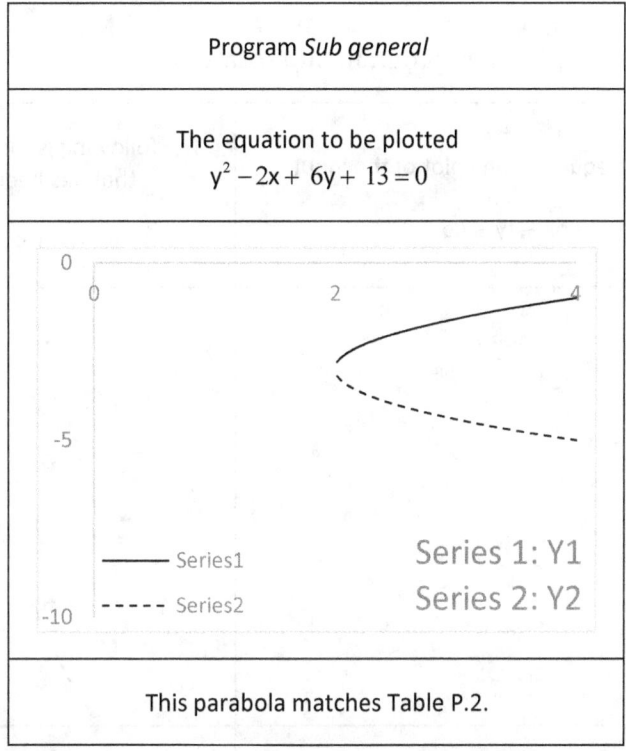

| Program *Sub general* |
| :---: |
| The equation to be plotted<br>$y^2 - 2x + 6y + 13 = 0$ |
| This parabola matches Table P.2. |

Series 1: Y1
Series 2: Y2

The second program made the plots in section *Rotating Conic Equations*. The polynomial to be rotated is: $A_{in}x^2 + B_{in}xy + C_{in}y^2 + D_{in}x + E_{in}y + F_{in} = 0$. The program computes the new coefficients and then calls **Sub equations** to compute y1 and y2. The following is the listing.

```
Sub rotate()
  Ain = Cells(5, 1): Bin = Cells(6, 1): Cin = Cells(7, 1)
  Din = Cells(8, 1): Ein = Cells(9, 1): Fin = Cells(10, 1)
  Angle = Cells(11, 1) * 4 * Atn(1) / 180

  cosT = Cos(Angle): sinT = Sin(Angle)
  A = Ain * cosT ^ 2 + Cin * sinT ^ 2 - Bin * cosT * sinT
  B = 2 * sinT * cosT * (Ain - Cin) + Bin * (cosT ^ 2 - sinT ^ 2)
  C = Ain * sinT ^ 2 + Cin * cosT ^ 2 + Bin * sinT * cosT
  D = Din * cosT - Ein * sinT
  E = Din * sinT + Ein * cosT
  F = Fin

  Call equations(A, B, C, D, E, F)
  Cells(5, 2) = A: Cells(6, 2) = B: Cells(7, 2) = C
  Cells(8, 2) = D: Cells(9, 2) = E: Cells(10, 2) = F
End Sub  ' rotate

Sub equations(A, B, C, D, E, F)

  Same as in Program Sub general

End Sub
```

• Read the input coefficients and the angle.

$$A_{in}x^2 + B_{in}xy + C_{in}y^2 + D_{in}x + E_{in}y + F_{in} = 0$$

• Compute the coefficients for the rotated conic. Note: Bin has been added to *Equation coefficients*.

$$A = A_{in}\cos^2\theta + C_{in}\sin^2\theta - B_{in}\sin\theta\cos\theta$$
$$B = 2\sin\theta\cos\theta(A_{in} - C_{in}) + B_{in}(\cos^2\theta - \sin^2\theta)$$
$$C = A_{in}\sin^2\theta + C_{in}\cos^2\theta + B_{in}\sin\theta\cos\theta$$
$$D = D_{in}\cos\theta - E_{in}\sin\theta$$
$$E = D_{in}\sin\theta + E_{in}\cos\theta$$
$$F = F_{in}$$

The following is an example using program **Sub rotate**.

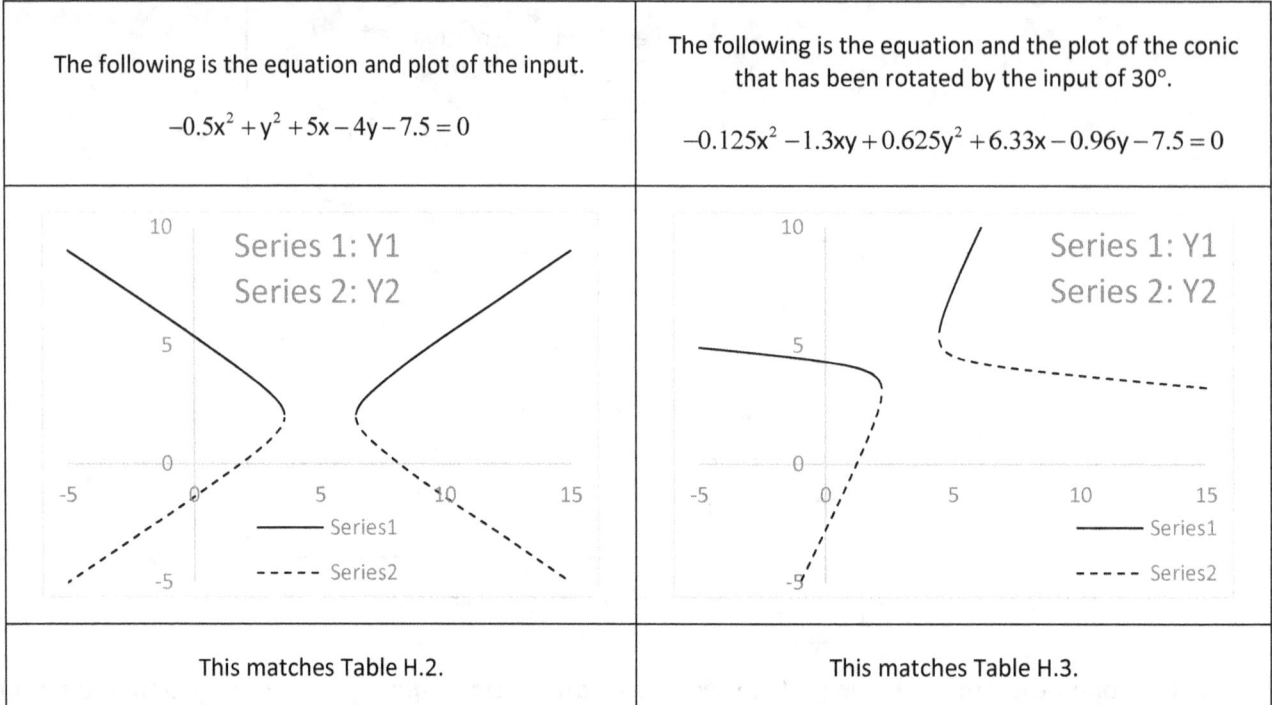

| The following is the equation and plot of the input. $$-0.5x^2 + y^2 + 5x - 4y - 7.5 = 0$$ | The following is the equation and the plot of the conic that has been rotated by the input of 30°. $$-0.125x^2 - 1.3xy + 0.625y^2 + 6.33x - 0.96y - 7.5 = 0$$ |
|---|---|
| This matches Table H.2. | This matches Table H.3. |

The third program generated the data for section *Unrotating Conic Equations*. For the equation $A_{in}x^2 + B_{in}xy + C_{in}y^2 + D_{in}x + E_{in}y + F_{in} = 0$, the program computes the angle that will remove the xy term, computes new coefficients, and then computes y1 and y2. The following is the listing of program Sub rotate_back.

<table>
<tr>
<td>

```
Sub rotate_back()
  Ain = Cells(5, 1): Bin = Cells(6, 1): Cin = Cells(7, 1)
  Din = Cells(8, 1): Ein = Cells(9, 1): Fin = Cells(10, 1)

  If Abs(Ain - Cin) < 0.01 Then
    Angle = 3.14159 / 4
  Else
    Angle = 0.5 * Atn(Bin / (Ain - Cin))
  End If

  cosT = Cos(Angle): sinT = Sin(Angle)
  A = Ain * cosT ^ 2 + Cin * sinT ^ 2 + Bin * cosT * sinT
  B = 2 * sinT * cosT * (Cin - Ain) + Bin * (cosT ^ 2 - sinT ^ 2)
  C = Ain * sinT ^ 2 + Cin * cosT ^ 2 - Bin * sinT * cosT
  D = Din * cosT + Ein * sinT
  E = -Din * sinT + Ein * cosT
  F = Fin

  Call equations(A, B, C, D, E, F)
  Cells(5, 2) = A: Cells(6, 2) = B: Cells(7, 2) = C
  Cells(8, 2) = D: Cells(9, 2) = E: Cells(10, 2) = F
  Cells(11, 2) = Angle * 180 / (4 * Atn(1))
End Sub  ' rotate_back

Sub equations(A, B, C, D, E, F)

  Same as in Program Sub general

End Sub
```

</td>
<td>

- Read the input coefficients.

$$A_{in}x^2 + B_{in}xy + C_{in}y^2 + D_{in}x + E_{in}y + F_{in} = 0$$

- Compute the angle. $\theta = \dfrac{1}{2}\tan^{-1}\left(\dfrac{B_{in}}{A_{in}-C_{in}}\right)$

- Compute the coefficients using *Equation rotate_back*.

$$A = A_{in}\cos^2\theta + C_{in}\sin^2\theta + B_{in}\sin\theta\cos\theta$$

$$B = 2\sin\theta\cos\theta\left(C_{in}-A_{in}\right) + B_{in}\left(\cos^2\theta - \sin^2\theta\right)$$

$$C = A_{in}\sin^2\theta + C_{in}\cos^2\theta - B_{in}\sin\theta\cos\theta$$

$$D = D_{in}\cos\theta + E_{in}\sin\theta$$

$$E = -D_{in}\sin\theta + E_{in}\cos\theta$$

$$F = F$$

</td>
</tr>
</table>

The following is an example using program **Sub rotate_back**.

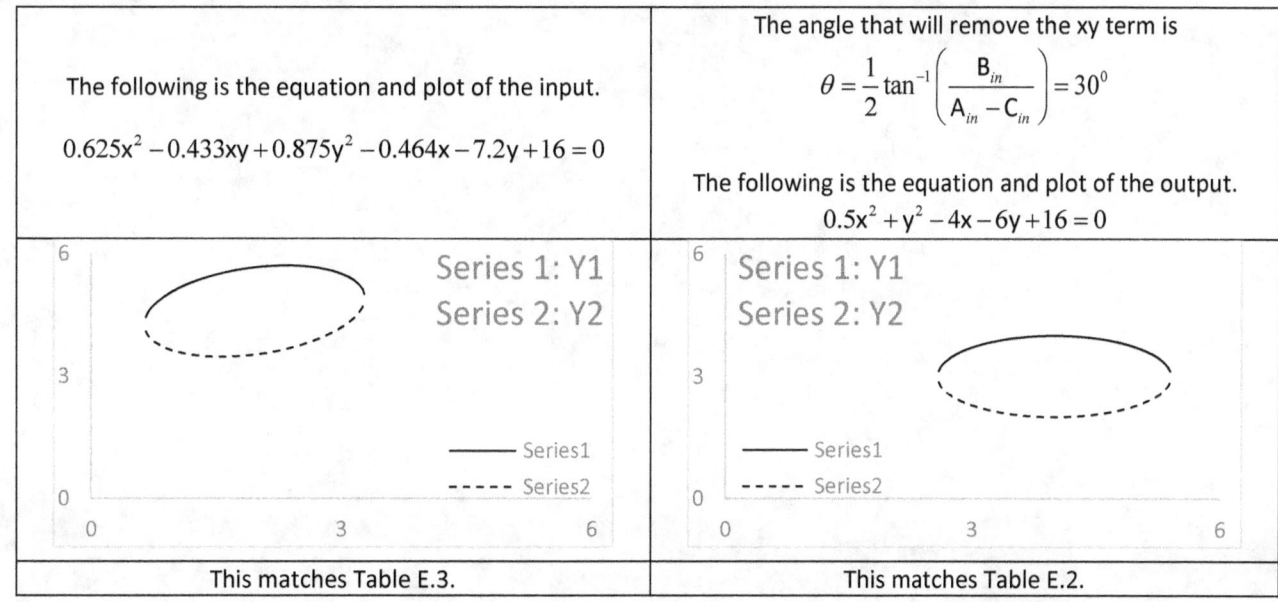

| The following is the equation and plot of the input. $$0.625x^2 - 0.433xy + 0.875y^2 - 0.464x - 7.2y + 16 = 0$$ | The angle that will remove the xy term is $$\theta = \frac{1}{2}\tan^{-1}\left(\frac{B_{in}}{A_{in}-C_{in}}\right) = 30^0$$ The following is the equation and plot of the output. $$0.5x^2 + y^2 - 4x - 6y + 16 = 0$$ |
|---|---|
| Series 1: Y1  Series 2: Y2  —— Series1  - - - - Series2 | Series 1: Y1  Series 2: Y2  —— Series1  - - - - Series2 |
| This matches Table E.3. | This matches Table E.2. |

# Chapter 8: Summary of Excel with VBA

This chapter is intended to stand-alone and summarize VBA for those involved in mathematics. On the VBA language that has already been presented, more detail is given.

## Section 8.1: The Programming Environment Between Excel and VBA

Open Excel and create a new workbook. Then **Save As:**

| For this discussion, save the workbook as indicated below. |
| --- |
| Filename:  First<br><br>Where:  Desktop<br><br>Type:  Excel Macro-Enabled Workbook (.xlsm) |
| When the file named First is reopened, *click* the Enable Macros button if required. |

On a tab near the bottom-left of the screen, the spreadsheet is called **Sheet1**. The + next to this tab is for adding more spreadsheets. For now, leave it at **Sheet1**.

The rows of the spreadsheets are given numbers. The columns are given letters. If you want to change the letters to numbers, see the following table.

| How to Switch the Columns from Letters to Numbers |
| --- |
| On a PC with Excel 2013 or 2016: File → Options → Formulas → R1C1 Reference Style.<br><br>On a Mac with Excel 2011: Excel → Preferences → General → R1C1 Reference Style.<br><br>If these instructions don't work on your system, Google<br>    *How to switch to R1C1 reference style in Excel* [your version]. |

If the **Developer** tab is not shown, it must be activated. See the following table.

| How to Activate the Developer Tab |
| --- |
| On a PC with Excel 2013 or 2016: File → Options → Customize Ribbon → Developer.<br><br>On a Mac with Excel 2011: Excel → Preferences → Ribbon → Developer.<br><br>If these instructions don't work on your system, Google<br>    *How to activate the Developer tab in Excel* [your version]. |

The following is a step-by-step discussion on how to write a VBA program. Click the **Developer** tab. On its ribbon, click **Editor** (or **Visual Basic**). Two things happen.

| 1)   The *Project* Window Opens. | 2) The Category *Run* Appears in the Menu Bar. |
|---|---|
| <br>**Project**<br><br>**VBA Project (First.xlsm)**<br>• Sheet1 (Sheet1)<br>• ThisWorkbook<br><br>Sheet1(Sheet1) and This Workbook are names of two windows that can contain VBA code. They are currently empty. | Selecting Run → Run Macro, opens the Macros window. It lists the names of programs that can be run right now.<br><br>**Macros**<br><br>*Initially Empty* |

## Writing a Program

1) In the **Project** window, double-click **ThisWorkbook**. The code window will open.

2) At the cursor in the empty window, enter the following five lines of VBA code.

| First.xlsm: ThisWorkbook (Code) | |
|---|---|
| Code | Comments |
| Sub one()<br>  Cells(1, 1) = 4 * Atn(1)<br>  pi = Cells(1, 1)<br>  Cells(1, 3) = 2 * pi<br>End Sub | •The name of the program is *one*. The empty parentheses are necessary. More in section 8.7.<br><br>•The *Cells...* statement writes the value of $\pi = 4*\tan^{-1}(1)$ into spreadsheet cell row 1, column 1.<br><br>•The *pi=...* statement reads the value of $\pi$ from cell row 1, column 1.<br><br>•This *Cells...* statement writes the value of $2*\pi$ into cell row 1, column 3. |

3)  In the top menu bar, File→Save and close the window.

4)  Select Run→Run Macro. This opens the Macros window. It will show an item called **ThisWorkbook.one**. Click on this item and then select **Run**.

5)  On the spreadsheet, see the following:

- The number $\pi$ is written in the cell row 1, column 1.
- The number $2\pi$ is written in row 1, column 3.

6)  Prepare for the next run. Clear the spreadsheet by selecting (highlighting) the data and then clicking Edit→Clear→All. Selecting is done by holding-down-the-clicker and dragging.

Via the + at the bottom of the spreadsheet, add **Sheet2**. Then click **Editor (Visual Basic)**. This opens the **Project** window. Observe that **Sheet2(Sheet2)** has been added. The windows **Sheet1(Sheet1)** and **Sheet2(Sheet2)** are currently empty.

| Project |
|---|
| <u>VBA Project (First.xlsm)</u> |
| Sheet1(Sheet1) |
| Sheet2(Sheet2) |
| ThisWorkbook |

Select **Run→Run Macro**. In the **Macros** window, click **ThisWorkbook.one** and Run. This time, $\pi$ and $2\pi$ are written on **Sheet 2**. This shows that the program named **one**, in ThisWorkbook, reads and writes to the open (visible) spreadsheet.

## Adding More Code Windows

We will now add another code window. If the **Project** window is not open, click **Editor**. Then **Insert→Module**. The **Project** window now contains four code windows.

| Project |
|---|
| <u>VBA Project (First.xlsm)</u> |
| Module1 |
| Sheet1(Sheet1) |
| Sheet2(Sheet2) |
| ThisWorkbook |

We will now copy the program named **one** from **ThisWorkbook** to the other three code windows.

- Double-click ThisWorkbook, Edit→Select All→Copy, and close the window.
- Double-click Module1, Edit→Paste, File→Save, and close the window.
- Double-click Sheet1, Edit→Paste, File→Save, and close the window.
- Double-click Sheet2, Edit→Paste, File→Save, and close the window.

Now, select **Run→Run Macro** to open the **Macros** window.

| Macros |
|---|
| one |
| Sheet1.one |
| Sheet2.one |
| ThisWorkbook.one |

We can now run four programs. We have shown that **ThisWorkbook.one** uses the open (visible) spreadsheet. Program **one**, which is in **Module1**, also uses the open spreadsheet. Program **Sheet1.one** uses only spreadsheet 1. Program **Sheet2.one** uses only spreadsheet 2.

- More modules can be added. Each **Module** $j$ uses the open spreadsheet.
- More sheets can be added. Each **Sheet** $j$ uses spreadsheet $j$.

## Notes

• By using ThisWorkbook and the Module*j* code windows, a program can run many cases, changing only the spreadsheets between cases.

• Code windows can contain more than one program.

• Programs in different code windows communicate with each other through the spreadsheets.

• Programs in the same window can also communicate with each other through the spreadsheets. But as will be shown in section 8.7, they can also communicate with each other directly.

• Programs read and write to specific cells. Cutting and pasting data from these cells to others can keep the specific cells free for subsequent runs.

• An attractive feature of the spreadsheet is that, after a run, clarifying comments can be manually added.

• To select two or more areas on the spreadsheet:
  • First area, hold-down-clicker and drag;
  • Subsequent areas, hold-down-command-key while holding-down-clicker and dragging, or hold-down-ctrl-key while holding-down-clicker and dragging.

• All code windows can be given usernames. The Properties window is where this is done. To access this window from the menu bar, View→Properties window.

• On the menu bar is the command View→Code. This opens a code window just like double-clicking on its name.

• The Project window contains all of the files that are open. This facilitates things like copying from one file to another.

• It is good practice to close code windows before selecting Run→Run Macro.

### Printing the Spreadsheet Data

Spreadsheet data can be printed directly from Excel. Data can also be pasted into Word and PowerPoint. Tip: Paste Options→Paste Special→Bitmap (or PDF).

## Other Types of Read-Write Statements

These statements allow any program in any window to read from and write to any spreadsheet. They are not used in this book. The following table shows two programs in the **Sheet1** code window. Program **two** uses these alternate read-write statements.

| Programs one and two are in the Sheet 1 Code Window |
|---|

```
Sub one()
  pi = 4 * Atn(1)                          ' pi = л

  Cells(1, 1) = pi                         ' This statement writes л to sheet 1
  ppp = Cells(1, 1)                        ' This statement reads л from sheet 1

End Sub


Sub two()
  pi = 4 * Atn(1)                          ' pi = л

  Sheets(2).Cells(1, 1) = pi               ' This statement writes л to sheet 2
  ppp = Sheets(2).Cells(1, 1)              ' This statement reads л from sheet 2

  Worksheets("Sheet3").Cells(1, 1) = ppp   ' This statement writes л to sheet 3
  eee = Worksheets("Sheet3").Cells(1, 1)   ' This statement reads л from sheet 3

  Cells(1, 1) = eee                        ' This statement writes л to sheet 1

End Sub
```

### Notes

• The prefixes Worksheets("Sheet *j* ") and Sheets( *j* ) direct the Cells statement to Sheet *j*, only if Sheet *j* has been added to the Project Window.

• An apostrophe causes the rest of the line to be ignored. Use this to add comments to the code.

## Section 8.2: Making Graphs

This section describes three programs and how to graph their outputs.

### Program graph1

This program reads values of y from a spreadsheet, computes $y^2$ and $y^3$, and then writes these back onto the same spreadsheet. This involves repeating a series of calculations a predetermined number of times. It will use the VBA statements for looping known as **For ... Next**.

| Syntax for Looping via For ... Next |
|---|
| For index = start-value To end-value Step step-value<br>    {statements in the loop}<br>Next index |
| Notes:<br>• index is a variable (integer or floating point).<br>• start-value, end-value, and step-value may be integers or floating points and positive or negative.<br>• If Step is not used, Step 1 is the default. |

The following is a listing of the program and its output.

```
graph1.xlsm: ThisWorkbook (Code)


Sub graph1()
For i = 1 To 9
   y =  Cells( i, 1 )
   Cells( i, 2 ) = y
   Cells( i, 3 ) = y ^ 2
   Cells( i, 4 ) = y ^ 3
Next i
End Sub
```

### The Spreadsheet from Program graph1

| Y (User Input)<br>Column 1 | Y (Series 1)<br>Column 2 | $Y^2$ (Series 2)<br>Column 3 | $Y^3$ (Series 3)<br>Column 4 |
|---|---|---|---|
| -1.5 | -1.5 | 2.25 | -3.375 |
| -1.2 | -1.2 | 1.44 | -1.728 |
| -0.8 | -0.8 | 0.64 | -0.512 |
| -0.4 | -0.4 | 0.16 | -0.064 |
| 0 | 0 | 0 | 0 |
| 0.4 | 0.4 | 0.16 | 0.064 |
| 0.8 | 0.8 | 0.64 | 0.512 |
| 1.2 | 1.2 | 1.44 | 1.728 |
| 1.5 | 1.5 | 2.25 | 3.375 |

Values of y are input by the user in column 1. The program reads these values, and then writes them into column 2. This column will be called **Series 1**. For each value of y, the program then computes $y^2$ and $y^3$, and writes these into columns 3 and 4.

The following is a plot of the output.

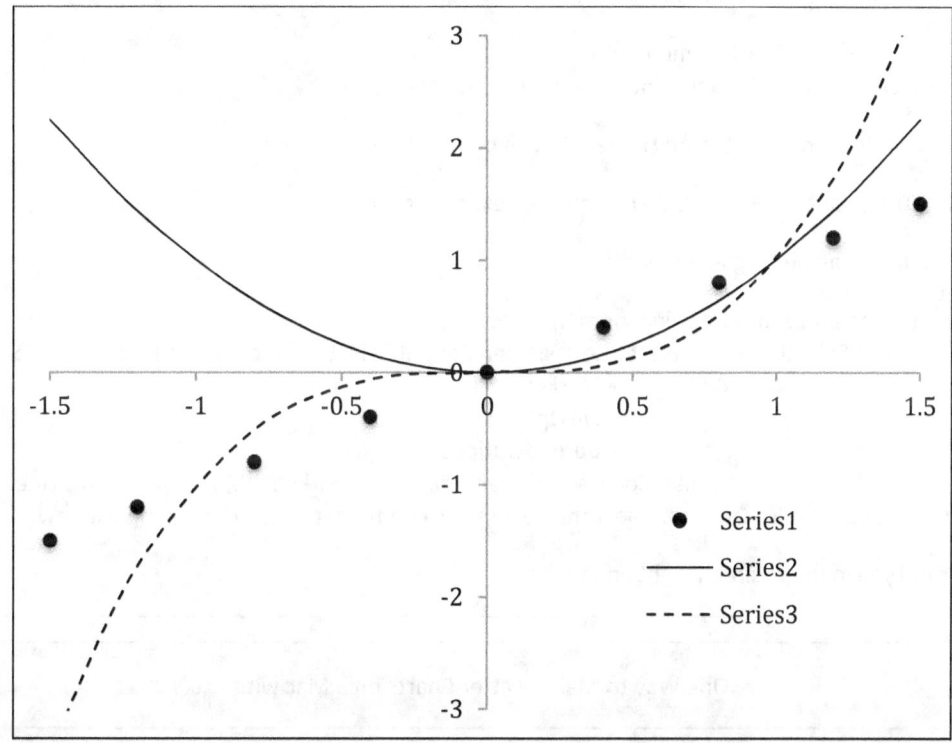

Excel refers to this plot as a **Scatter** chart or X-Y chart. Of the columns of data selected to be plotted, the horizontal axis is on the left. All of the other columns are for the vertical axis, and are called **Series 1, 2**, and so on. Each series can have its own format, for example, solid line, dashed line, or markers.

If the points are to be connected, the horizontal axis data must be monotonic. Each axis has its own scaling (minimum, maximum, increment). The scales for either or both axes can be logarithmic.

The following two tables show how to make a **Scatter** chart. One is for a PC, and the other is for a Mac.

---

### One Way to Make Scatter Charts on a PC with Excel 2013 and 2016

• Select (highlight) the data to be plotted. Select the Insert Tab→ Charts→ Scatter→ Scatter with Smooth Lines.

• Click the chart to bring up a ribbon with tabs for Design and Format.

(1) Design. Under Add Chart Elements:
   (A) Axes. Options include range, increments, and tick marks.

   (B) Chart Title. Can be edited on Home tab. Can be moved by dragging.

   (C) Axis Titles. Options→Size and Properties→Text direction.

   (D) Gridlines. Can be toggled on or off.

(2) Format. Each series may be a line, or markers, or markers with a line.
   (A) In Current Selection area, select a series. Then Format Selection to open the Format Data Series window.
Under the paint bucket: (a) Marker. • Marker Options.
                                     • Fill Options.
                                     • Border Options.
              (b) Line. Color, weight, and dash. For markers: (b) Line. Line→No Line.
If necessary, under Series Options, select the next series and repeat. Finally, close the window.

   (B) Size. Type in the desired height and width.

---

### One Way to Make Scatter Charts on a Mac with Excel 2011

• Select (highlight) the data to be plotted. Select the Charts Tab→ Scatter→ Smooth Lined Scatter.

• Click the chart to bring up a ribbon with tabs for Chart Layout and Format.

(1) Chart Layout.
   (A) Axes. Options include scale for range, increments, and tick marks.

   (B) Chart Title. Can be edited on Home tab. Can be moved by dragging.

   (C) Axis Titles. Options→Textbox→Text direction.

   (D) Gridlines. Can be toggled on or off.

(2) Format. Each series may be a line, or markers, or markers with a line.
   (A) In Current Selection area, select a series. Then Format Selection to open the Format Data Series window.
              (a) Marker Style. Options.
              (b) Marker Fill. Options.
              (c) Marker Line. Options are for the border of each marker.
              (d) Line. Color, weight, and dash. For markers: (d) Line. Color→No Line.
Click OK to close the Format Data Series window. In Current Selection area, select next series and repeat.

   (B) Size. Type in the desired height and width.

## Program graph2

The block diagram and plotted output of the program are shown in the following figure.

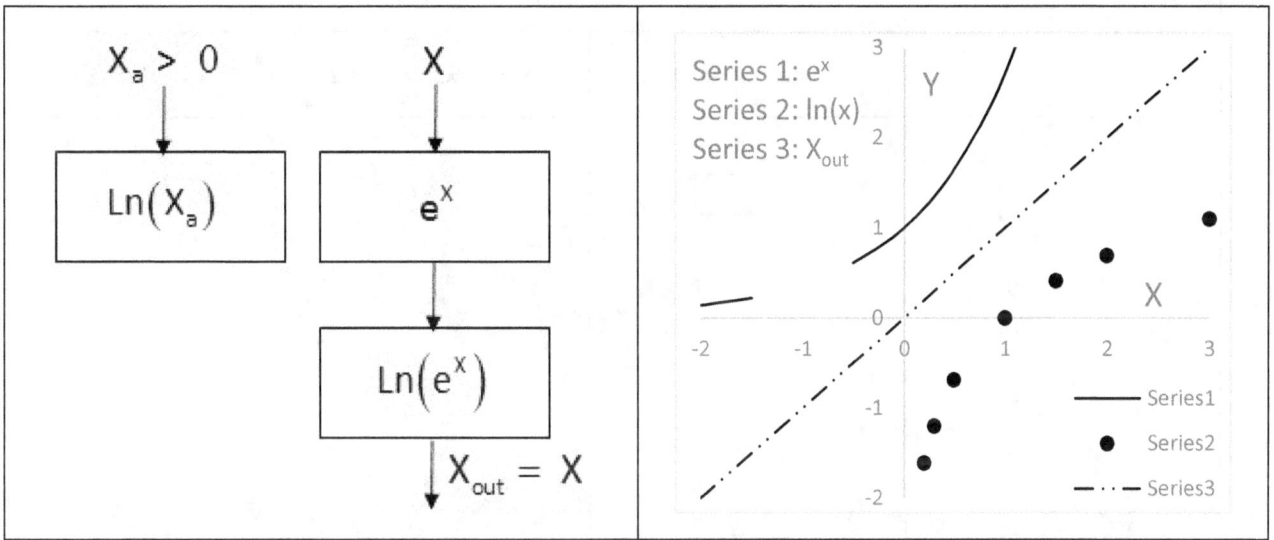

Each output variable is computed from its own set of horizontal axis values. This program shows how these variables are printed on the spreadsheet so they can share a single horizontal axis. Note that Series 1 has two segments. That will be discussed later. The following table shows the program.

| The Program Sub graph2 | |
| --- | --- |
| Option Base 1<br>Sub graph2()<br>  j = 1<br>  For x = -2 To 3 Step 0.5<br>    Y = Exp(x)<br>    Cells(j, 1) = x<br>    Cells(j, 2) = Application.Round(Y, 2)<br>    j = j + 1<br>  Next x<br>  Xa = Array(3, 0.1, 2, 0.2, 1.5, 0.3, 1, 0.5)<br>  For j = 1 To 8<br>    Y = Log(Xa(j))<br>    Cells(j + 11, 1) = Xa(j)<br>    Cells(j + 11, 3) = Application.Round(Y, 2)<br>  Next j<br>  j = 1<br>  For x = -2 To 3 Step 0.5<br>    Xout = Log(Exp(x))<br>    Cells(j + 19, 1) = x<br>    Cells(j + 19, 4) = Application.Round(Xout, 2)<br>    j = j + 1<br>  Next x<br>End Sub | • Option Base 1 specifies that the first element in the Xa array is Xa(1).<br><br>• Syntax: Exp(x) is $e^x$ ; Log(x) is Ln(x); and Application.Round(Y, 2) rounds Y to two decimal places.<br><br>• Y = $e^x$ and is put in column 2 and called Series 1.<br><br><br>• Y = Ln(x) and is put in column 3 and called Series 2. This series gets the values of x from an Array function. In this way, x can be numbers with uneven intervals which is appropriate for the log function. Because these numbers are also out of order, Series 2 must be plotted using markers.<br><br>• Y = Ln(x) and is put in column 4 and called Series 3.<br><br>• Note that arithmetic can be done in the Cells statement. |

The output is shown in the following table. Each variable is offset so it can be formatted separately.

| The Spreadsheet from Program Sub graph2 | | | |
|---|---|---|---|
| X - axis | $Y = e^x$ : Series 1 | $Y = Ln(Xa)$: Series 2 | $Y = Ln(e^x)$: Series 3 |
| -2 | 0.14 | | |
| -1.5 | 0.22 | | |
| | | | |
| -0.5 | 0.61 | | |
| 0 | 1 | | |
| 0.5 | 1.65 | | |
| 1 | 2.72 | | |
| 1.5 | 4.48 | | |
| 2 | 7.39 | | |
| 2.5 | 12.18 | | |
| 3 | 20.09 | | |
| 3 | | 1.1 | |
| 0.1 | | -2.3 | |
| 2 | | 0.69 | |
| 0.2 | | -1.61 | |
| 1.5 | | 0.41 | |
| 0.3 | | -1.2 | |
| 1 | | 0 | |
| 0.5 | | -0.69 | |
| -2 | | | -2 |
| -1.5 | | | -1.5 |
| -1 | | | -1 |
| -0.5 | | | -0.5 |
| 0 | | | 0 |
| 0.5 | | | 0.5 |
| 1 | | | 1 |
| 1.5 | | | 1.5 |
| 2 | | | 2 |
| 2.5 | | | 2.5 |
| 3 | | | 3 |

Note: In Series 1, there is a blank space in the values. This data point was manually removed in order to illustrate plotter *pen-lifting*. That's why Series 1 has two segments.

## Program graph 3

Excel has another kind of chart, called a **Surface** chart. It is like a topographical map. Its three axes are called vertical, horizontal, and depth. The vertical axis data has numerical values, whereas the data for the other axes are bins or categories. But these categories can be numbers. If they correspond to actual values, the **Surface** chart becomes a true 3-D chart. This is demonstrated in the following program.

This program computes $f(i,j) = \sqrt{(i-2)^2 + (j-2)^2}$, for $i = 0\ to\ 4$ and $j = 0\ to\ 4$. The values of $f(i,j)$ are output to the spreadsheet in matrix form to be used for the vertical axis. The values for $i$ and $j$ are the categories for the depth and horizontal axes.

The following is a listing of the code and the spreadsheet output. The code illustrates the use of nested **For ... Next** loops, as well as **VBA** for computing the square-root.

---

**graph3.xlsm: ThisWorkbook (Code)**

```
Sub graph3()
For i = 0 To 4
  For j = 0 To 4
     f = Sqr( ( i − 2 )^2 + ( j − 2 )^2 )
     Cells( 2 + i, 2 + j ) = Application.Round( f, 2 )
     Cells ( 1, 2 + j ) = j
  Next j
  Cells( 2 + i, 1 ) = i
Next i
End Sub
```

---

### The Spreadsheet from Program graph3

| | Column 1 | Values of *j* Used as Horizontal Axis Categories | | | | |
| --- | --- | --- | --- | --- | --- | --- |
| | Column 1 | Column 2 | Column 3 | Column 4 | Column 5 | Column 6 |
| Row 1 | | 0 | 1 | 2 | 3 | 4 |
| Values of *i* | 0 | 2.83 | 2.24 | 2 | 2.24 | 2.83 |
| Used as | 1 | 2.24 | 1.41 | 1 | 1.41 | 2.24 |
| Depth Axis | 2 | 2.00 | 1.00 | 0 | 1.00 | 2.00 |
| Categories | 3 | 2.24 | 1.41 | 1 | 1.41 | 2.24 |
| | 4 | 2.83 | 2.24 | 2 | 2.24 | 2.83 |

The following is a **Surface** chart of this data. The spreadsheet highlighted for plotting is cell(1,1) to cell(6,6). Note that cell(1,1) is intentionally blank.

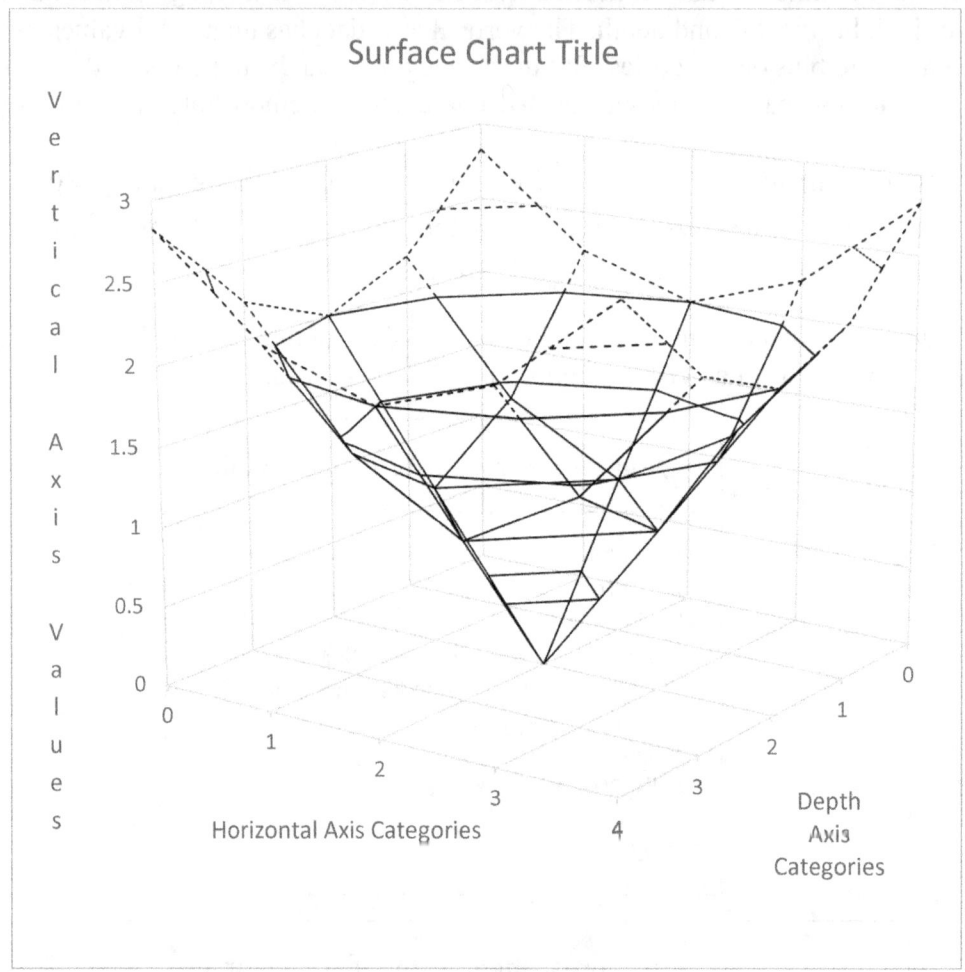

The following two tables show how to make a **Surface** chart. One is for the PC. The other is for the Mac.

---

### One Way to Make Surface Charts on a PC with Excel 2013 and 2016

Select (highlight) the data to be plotted. Select the Insert Tab→ pentagon icon→ Surface→ Wireframe 3-D Surface. Click the chart to bring up the Chart Tools ribbon with: (1) Format tab; (2) Design tab.

(1) Format tab with: (A) Size; (B) Shape Styles with Shape Outline.
  (A) I recommend sizing as the first step. Click the chart→ Format tab→ Size.

  (B) Click the legend. A second click selects one series. On the Format tab: Shape Outline for color, weight, dash. If deleted, legend can be restored using Add Chart Elements on Design tab.

(2) Design tab→ Add Chart Elements. In its list: (A) Axes; (B) Chart Title; (C) Axis Titles; (D) Gridlines; (E) Legend.
  (A) Axes. Toggle horizontal, vertical, and depth. Options→ Format Axis window→ Axis Options. In its list:
  • Vertical (Value) Axis→ three-bars icon→ Axis Options: min/max/major units. Close window.
  • Depth (Series) Axis→ three-bars icon→Options: series in reverse order. Close window.
  • Plot Area→ Format Plot Area window→ pentagon icon→ 3-D rotation. Close window.

  (B) Chart Title. This adds a textbox, which can be edited from the Home tab. It can also be moved.

  (C) Axis Titles. Toggle horizontal, vertical and depth textboxes. Options→Format Axis Title window→Title Options→ Select an axis title→Text Options→Textbox icon→Text box→Text Direction. Close window.

  (D) Gridlines. Horizontal, vertical and depth can be toggled.

---

### One Way to Make Surface Charts on a Mac with Excel 2011

Select (highlight) the data to be plotted. Select the Charts tab→ Other→ Surface→ Wireframe 3-D Surface. Click on the chart. This brings up a ribbon with: (1) Format tab; (2) Chart Layout tab.

(1) Format Tab with: (A) Size; (B) Line.
  (A) I recommend sizing as the first step. Click the chart→ Size.

  (B) Click the legend. A second click selects one series. On the Format tab: Line for color, weight, dash. If deleted, Legend can be restored using Chart Layout tab.

(2) Chart Layout with: (A) Axes; (B) Chart Title; (C) Axis Titles; (D) Gridlines; (E) Rotation; (F) Legend. Click on the chart.
  (A) Axes→horizontal, vertical, and depth can be toggled.
    • For vertical axis. Options→ Format Axis window→ Scale: min, max, major unit. Click OK.
    • For depth axis. Options→ Format Axis window→ Scale. If desired, select series in reverse order. Click OK.

  (B) Chart Title. This adds a textbox, which can be edited from the Home tab. It can also be moved.

  (C) Axis Titles. Toggle horizontal, vertical, depth textboxes. Options→ Format Title window→Textbox→Text Direction.

  (D) Gridlines. Horizontal, vertical and depth can be toggled.

  (E) 3-D Rotation.

## Printing a Chart

A chart can be printed directly from Excel. It can also be pasted into Word and PowerPoint. One way to paste a chart into Word is shown in the following table.

---

- Click on the chart and copy.
- Paste into PowerPoint as a graphic object, and resize, reformat, and annotate if desired.
- Copy from PowerPoint.
- Paste into Word via Paste Special → PDF or Paste Special → Bitmap.

---

Another way to paste into Word is

$$\text{Paste} \rightarrow \text{Paste Special} \rightarrow \text{Graphic Object.}$$

This allows you to alter the appearance of the chart in Word.

# Section 8.3: Arrays and Data Types

Variables that are assigned multiple values are called arrays. Indices are used to refer to a particular value in an array. For example, if five values are assigned to variable X, then X appears in code as X(i), and i can range from 1 to 5. Then the statement Y = X(3) assigns to Y the value in the X array for i = 3.

Array variables can have more than one index. Consider the variable beta(i, j, k). If i ranges from 1 to 2, j ranges from 1 to 4, and k ranges from 1 to 5, then beta can have forty values.

## The Dim Statement

The program compiler (interpreter) must know in advance, how much space in memory to set aside for each array variable. This is done via the Dim statement. All arrays must be *sized* in a Dim statement. The one exception will be discussed later. Here is an example of a Dim statement.

$$\text{Dim } u(9,10), v(5)$$

This statement allocates ninety spaces for u, and five spaces for v.

• The Dim statement can allocate more spaces than are actually needed.

• There may be multiple Dim statements.

• A Dim statement can be placed anywhere in the code, as long as it appears before each of its variables are used.

## Another Use of the Dim Statement

When the VBA statement Option Explicit is the first line of the code, all variables used in the program must appear in a Dim statement. With this statement in the code, the names of all the variables are easily seen in the Dim statements.

## A Third Use of the Dim Statement

All variables in a program are classified according to their *type*. *Types* include integer, floating point, character, and logical. When it is necessary to declare that a variable is a certain type, this is done in a Dim statement.

VBA has a so-called *smart* type. In fact, it's the default type for all the variables. It is illustrated in the following two programs.

| Program Illustrating the Variant Data Type | |
|---|---|
| *Notes:* <br> • If the type is not specified, type *variant* is assumed. Hence, in this program, variable b is a *variant*. <br> • The data type of a variant automatically adapts during the program. <br> • *Variant* is the only data type that is used in any program in this book. | |
| **Code** | **Comments** |
| Sub variant_1() <br>  b = " first " <br>    Cells(1, 1) = b <br>  b = 11 <br>    Cells(2, 1) = b <br>  b = 77.5 <br>    Cells(3, 1) = b <br>  b = Array(" second ", 33, " third ", 3.14159) <br>    For I = 0 To 3 <br>      Cells(4, I + 1) = b(i) <br>    Next i <br>  b = 55 <br>    Cells(5, 1) = b <br>End Sub | • b starts as a character variable whose value is specified within quotes. <br> • Then, b automatically becomes an integer. <br><br> • b becomes a floating point number. <br><br> • b becomes an array of dimension *four*. <br>    • By default [Option Base 0], first item is b(0). <br>    • When $I$ = 0,  Cells(4, $I$ + 1) = Cells(4, 1). <br><br> • b becomes a floating point or integer number. <br> (The array function has been overridden.) |

The following table shows its output.

| The Values of b Output by Program variant_1 | | | | |
|---|---|---|---|---|
| Row | Column 1 | Column 2 | Column 3 | Column 4 |
| 1 | first | | | |
| 2 | 11 | | | |
| 3 | 77.5 | | | |
| 4 | second | 33 | third | 3.14159 |
| 5 | 55 | | | |

The following program uses the variable b as an array. It produces the same output.

| Second Program Illustrating the Variant Data Type | |
|---|---|
| Sub variant_2() <br>Dim b(10) <br>  b(0) = " first " <br>    Cells(1, 1) = b(0) <br>  b(0) = 11 <br>    Cells(2, 1) = b(0) <br>  b(0) = 77.5 <br>    Cells(3, 1) = b(0) <br>  b(0) = "second": b(1) = 33: b(2) = "third": b(3) = 3.14159 <br>    For I = 0 To 3 <br>      Cells(4, I + 1) = b(i) <br>    Next i <br>  b(0) = 55 <br>    Cells(5, 1) = b(0) <br>End Sub | • The Dim statement has been added. <br><br> • b(0) starts as a character variable. <br><br> • b(0)  automatically becomes an integer. <br><br> • b(0)  becomes a floating point number. <br><br> • Data types in the **b** array can differ. <br><br> *Note:* Multiple statements on a line are separated by colons. <br><br> • b(0)  becomes a floating point or integer number. |

## Option Base

VBA has another option statement called **Base**. When the statement **Option Base 1** is at the top of the code, the indices for all arrays start at unity. If **Option Base 0** is at the top, the indices for all arrays start at zero. **Option Base 0** is the default. The following two programs illustrate this. They produce the same result.

| Program Sub one | Program Sub two |
|---|---|
| Sub one()<br>alpha = Array(-25, 25, "four", 0, -10)<br>  For i = 0 To 4<br>    Cells(1 ,  i + 1) = alpha( i )<br>  Next i<br>End Sub | Option Base 1<br>Sub two()<br>alpha = Array(-25, 25, "four", 0, -10)<br>  For i = 1 To 5<br>    Cells(1 ,  i) = alpha( i )<br>  Next i<br>End Sub |
| Note the *arithmetic* required in the Cells statement.<br>When i = 0 , Cells(1 , i + 1) = Cells( 1 , 1 ) | |

## Final Notes

• This is the exception in the **Dim** statement that was mentioned earlier. A variable given values by the **Array** function, may be declared in a **Dim** statement, but must not be dimensioned in a **Dim** statement. In other words, if X = **Array**(2,4,6,8), it is permissible to have the statement **Dim X**. It is not permissible to have the statement **Dim X(4)**.

• The **Array** function is a convenient way of assigning values to a one-dimensional variable. It is awkard to use for a variable with two or more dimensions. Also, it can only assign values in the program code. Values cannot be assigned from the spreadsheet.

• The storage space for a **variant** variable is twice that for a double precision variable. However, its numeric range is the same as double precision. From Lomax[6], use of the **variant** data type does slightly increase program run time.

• All of the variables in the programs in this book are **variant** type.

• When an option statement (**Explicit** or **Base**) is used, it applies to all of the programs in that code window.

---

[6] Lomax, P., *VB and VBA in a Nutshell* (Sebastopol, CA: O'Reilly & Associates, 1998).

# Section 8.4: Dynamic Arrays and Excel Array Functions

## Dynamic Arrays

The Dim statement sets the dimensions of arrays. The ReDim statement allows the dimensions to be established while the program is running. The following table demonstrates the ReDim statement.

| Syntax for the ReDim Statement | |
|---|---|
| •<br>•<br>Dim A( )<br>•<br>•<br>•<br>n=3<br>ReDim A( n, n )<br>•<br>• | • The empty parentheses declares A to be a dynamic array.<br><br><br><br>• Later in the program, the size of the A matrix becomes known.<br>• This statement dimensions the A matrix. |
| Notes: • The statement Dim A( n, n ) is not allowed.<br>　　　　• The ReDim statement can be used only in the same program wherein its Dim statement appears. | |

## Excel Array Functions

There are several Excel functions for real arrays or matrices. We will demonstrate three of them while solving linear algebraic equations.

- Minverse to invert a real matrix;
- MMult to multiply two real matrices;
- MDeterm to compute the determinant of a real matrix.

The rules for using these are as follows:

• The arrays that are input to Minverse, MMult and MDeterm must be dimensioned to their exact size.

• The arrays that are output by Minverse, MMult and MDeterm must not be sized in a Dim statement.

The following program illustrates the use of Minverse, MMult, MDeterm, and the ReDim statements. Note the use of Option Base 1.

# Program Sub lin_equation

This program solves this linear algebraic equation:

$$A * x = b \quad \text{where} \quad A = \begin{bmatrix} 1 & 0 & 1 \\ 2 & 3 & 1 \\ 3 & 5 & 2 \end{bmatrix} \quad \text{and} \quad b = \begin{bmatrix} 4 \\ 8 \\ 6 \end{bmatrix}$$

The program finds x from this equation:

$$x = A^{-1} * b$$

The matrices and their size are input from the spreadsheet.

| Program Sub lin_equation | |
|---|---|
| Code | Comments |
| Option Base 1<br>Sub lin_equation()<br>  Dim A(), b()<br>    N = Cells(1, 1)<br>  ReDim A(N, N), b(N, 1)<br>  For i = 1 To N<br>    For j = 1 To N<br>      A(i, j) = Cells(i + 1, j)<br>    Next j<br>      b(i, 1) = Cells(i + 1, N + 1)<br>  Next i<br>    Ainv = Application.MInverse(A)<br>    x = Application.MMult(Ainv, b)<br>    bcheck = Application.MMult(A, x)<br>  For i = 1 To N<br>    Cells(i + N + 3, 1) = x(i, 1)<br>    Cells(i + N + 3, 3) = bcheck(i, 1)<br>  Next i<br>    deter = Application.MDeterm(A)<br>    Cells(N + 4, 5) = deter<br>End Sub | • Declare A and b to be dynamic matrices.<br>• N is the size of the A matrix.<br>• A and b can now be sized exactly (note that b *is sized as a column matrix*).<br><br><br><br>• A and b are read from the spreadsheet.<br><br><br>• Ainv = $A^{-1}$<br>• x = $A^{-1}$*b<br>• bcheck = A*x  This checks the answer.<br><br><br><br>• x , bcheck and deter are printed to the spreadsheet. |

The following table shows the spreadsheet.

| Spreadsheet for Program Sub lin_equation | | | | | | | |
|---|---|---|---|---|---|---|---|
| Input | | | | Output | | | |
| 3 | | | | | | | |
| 1 | 0 | 1 | 4 | 13 | | 4 | 2 |
| 2 | 3 | 1 | 8 | -3 | | 8 | |
| 3 | 5 | 2 | 6 | -9 | | 6 | |

# Section 8.5: Functions

Section 8.4 introduced the functions Minverse, MMult, and MDeterm. More functions are defined in this section. There are three types of functions:

- Excel functions
- VBA functions
- User-defined functions

All of Excel's functions are accessible to VBA programs. If a VBA function is the same as an Excel function, the VBA function must be used.

- The syntax for an Excel function is        y = Application.name(argument list)
- The syntax for a VBA function is            y = name(argument list)
- The syntax for a User-defined function is    y = name(argument list)

The following four tables list common Excel and VBA functions.

| Trig and Math Functions | |
|---|---|
| $y = \sin(\theta)$ with $\theta$ in radians | $y = \text{Sin}(x)$ |
| $y = \cos(\theta)$ with $\theta$ in radians | $y = \text{Cos}(x)$ |
| $y = \tan(\theta)$ with $\theta$ in radians | $y = \text{Tan}(x)$ |
| $\theta = \sin^{-1}(x)$ with $\theta$ in radians | $y = \text{Application.Asin}(x)$ |
| $\theta = \cos^{-1}(x)$ with $\theta$ in radians | $y = \text{Application.Acos}(x)$ |
| $\theta = \tan^{-1}(x)$ with $\theta$ in radians | $y = \text{Atn}(x)$ |
| $\theta = \tan^{-1}(x2/x1)$ with $\theta$ in radians | $y = \text{Application.Atan2}(x1, x2)$ |
| $y$ = square root of $x$ | $y = \text{Sqr}(x)$ |
| $y = e^x$ | $y = \text{Exp}(x)$ |
| $y = \log x$ to base 10 | $y = \text{Application.Log10}(x)$ |
| $y$ = natural log of $x$ | $y = \text{Log}(x)$ |
| $y = \log x$ to base $a$ | $y = \text{Application.Log}(x, a)$ |
| $y$ = absolute value of $x$ | $y = \text{Abs}(x)$ |
| $y$ = sign of $x$ (+1, 0 or -1) | $y = \text{Sgn}(x)$ |

| Functions That Place the Decimal Point | |
|---|---|
| y = integer by truncating x after the decimal point | y = Fix(x) |
| y = nearest integer to x in negative direction | y = Int(x) |
| y = nearest integer to x by rounding up or down | y = Round(x) |
| y = x rounded (up or down) to nd decimal places | y = Application.Round(x, nd) |
| y = x rounded up to nd decimal places | y = Application.RoundUp(x, nd) |
| y = x truncated to nd decimal places | y = Application.RoundDown(x, nd) |
| *Note: nd* may be positive, zero, or negative. | |

| Array Functions | |
|---|---|
| •Note: For Excel functions, their input arrays must be dimensioned exactly, and the output array must not be dimensioned (see section 8.3).<br>•By default, all array indices start at 0. To change to 1, use Option Base 1. | |
| y = maximum value of a general array | y = Application.Max(x) |
| y = minimum value of a general array | y = Application.Min(x) |
| y = inverse of x matrix | y = Application.Minverse*(x) |
| y = product of two matrices | y = Application.MMult(x1, x2) |
| y = determinant of x matrix | y = Application.MDeterm(x) |
| y = transpose of x matrix | y = Application.Transpose(x) |
| y = list of numbers assigned to a row array | y = Array(x1, x2, x3,...., xn) |
| * Minverse returns a scalar when its input is a 1x1 matrix. | |

| Complex Functions in Excel | |
| --- | --- |
| y = (a + $j$ b) combined into a single complex number | y = Application.Complex(a, b) |
| y = real coefficient of a complex number | y = Application.ImReal(x) |
| y = imaginary coefficient of a complex number | y = Application.Imaginary(x) |
| y = magnitude of a complex number | y = Application.ImAbs(x) |
| y = angle (in radians) of a complex number | y = Application.ImArgument(x) |
| y = quotient of two complex numbers | y = Application.ImDiv(x, z) |
| y = product of two or more complex numbers | y = Application.ImProduct(x, z, w) |
| y = sum of two or more complex numbers | y = Application.ImSum(x, z, w) |
| y = difference of two complex numbers | y = Application.ImSub(x, z) |
| y = complex number raised to an integer power | y = Application.ImPower(x, n) |
| y = $e^x$ where x is a complex number | y = Application.ImExp(x) |

The following table shows the syntax for a user-defined function.

| Syntax for a User-Defined Function |
|---|
| Sub main()<br><br>    •<br>    •<br>  y = *name*(argument list)<br>    •<br>    •<br><br>End Sub<br><br>Function *name*(argument list)<br>    •<br>    •<br>  Exit Function    '  Optional<br>    •<br>    •<br>  *name* = expression<br>End Function |

The following program illustrates a user-defined function. The function changes the dimensions of the angle named slant from radians to degrees.

| Sub example | |
|---|---|
| Code | Comments |
| Sub example()<br><br>  slant = 0.3<br>  slant = radTOdeg(slant)<br>  Cells(1, 1) = slant<br><br>End Sub<br><br>Function radTOdeg(slant)<br><br>  radTOdeg = slant * 180 / (4 * Atn(1))<br><br>End Function | • Note that the name of the angle does not change.<br><br>• $\pi = 4*\tan^{-1}(1)$ |

# Section 8.6:

## Looping and Branching,
## and
## the Operators (Comparison, Logical, and Arithmetic)

This section is divided into the following parts:
- Part A: Looping a definite number of times via For...Next
- Part B: Looping an indefinite number of times via Do...Loop
- Part C: Branching according to the *true/false* result of a condition test If-Then-Else
- Part D: Branching by comparing values using the Select Case statement
- Part E: The operators used in the following:
    - Arithmetic (addition, subtraction, etc.)
    - Condition tests (comparison and logical)

## • Part A: Looping via For...Next

Looping a definite number of times.

| Syntax for For...Next |
|---|
| For index = start-value To end-value Step step-value<br>    {statements in the loop that may include the Exit For statement}<br>Next index<br><br>where: 1. index is an integer or floating point variable.<br>    2. start-value, end-value, and step-value may be integer or floating point and positive or negative.<br>    3. Step is optional. Step 1 is the default.<br>    4. If end-value is less then start-value, execution transfers to after the For...Next loop.<br>    5. Exit For is a way to exit the loop before completion. It causes the program to branch<br>        to the statement after Next index. It is useful for program checkout. Exit For is optional. |

## • Part B: Looping via Do...Loop

The **Do...Loop** statement repeatedly executes a block of code *while* or *until* some condition is satisfied. Part E of this section discusses these condition tests. **Do...Loop** has five types.

• **Do While...Loop**: This executes the statements in the loop while some condition is true, and the condition is tested <u>before</u> the loop starts.

| Syntax for Do While...Loop |
| --- |
| Do While { *condition test* }<br>    {statements that may include Exit Do}<br>Loop<br><br>*Note:*<br>• If the *condition* test fails the first time, the statements in the loop will not be executed at all.<br>• Exit Do is optional. |

The following program demonstrates the **Do While...Loop**. This program estimates a computer's precision. The program determines the smallest difference between two numbers that the host computer can detect. Starting with an initial value of **epsilon = 1**, the program repeatedly divides **epsilon** by 2 until (1 + epsilon = 1).

| The Machine Epsilon Test Using a Do While...Loop | |
| --- | --- |
| Sub epsilon_test()<br>  epsilon = 1<br>  Do While ( 1 + epsilon ) > 1<br>    epsilon = epsilon / 2<br>  Loop<br>  epsilon = 2 * epsilon<br>End Sub | *Important Note:* This program shows a lesson about using While. Since the test fails inside the Do...Loop, an adjustment must be made outside the loop. That's the reason for the last line of code:<br>epsilon = 2 * epsilon.<br>*Conclusion:* A computer can detect a difference between numbers as small as this value of epsilon. For the Macbook OSX, and for the Hewlett-Packard 15 and ASUS tablet with Windows 8.1 and 10, the value of epsilon is $2^{-52}$. |

• **Do...Loop While**: This executes the statements in the loop while some condition is true, but the condition is tested <u>after</u> the loop is done.

| Syntax for Do...Loop While |
| --- |
| Do<br>    {statements that may include Exit Do}<br>Loop While { *condition test* }<br><br>*Note:*<br>• The statements in the loop will be executed at least once.<br>• Exit Do is optional. |

- **Part B (cont.):**

The word While can be replaced by Until. To make this replacement, adjustments must be made to the condition test. This will be shown in the sample program below.

• Do Until...Loop: This executes the statements in the loop until some condition is true, where the condition is tested <u>before</u> the loop starts.

| Syntax for Do Until...Loop |
|---|
| Do Until { *condition test* }<br>        {statements that may include Exit Do}<br>Loop<br>*Note:*<br>• if the *condition test* fails the first time, the statements in the loop will not be executed at all.<br>• Exit Do is optional. |

The following is a program demonstrating the **Do Until...Loop**. This is done by modifying the previous program that used the **Do While...Loop**. The condition test must be changed. The change is seen in the line of code Do Until (1 + epsilon) = 1.

| The Machine Epsilon Test Using a Do Until...Loop | |
|---|---|
| Sub epsilon_test()<br>  epsilon = 1<br>  Do Until ( 1 + epsilon ) = 1<br>    epsilon = epsilon / 2<br>  Loop<br>  epsilon = 2 * epsilon<br>End Sub | Important Note: This program demonstrates a lesson about using Until. Since the *condition test* fails inside the Do...Loop, an adjustment must be made outside the loop. That's the reason for the last line of the code:<br>                    epsilon = 2 * epsilon |

• Do...Loop Until: This executes the statements in the loop until some condition is true, but the condition is tested <u>after</u> the loop is done.

| Syntax for Do...Loop Until |
|---|
| Do<br>        {statements that may include Exit Do}<br>Loop Until { *condition test* }<br>*Note:*<br>• The statements in the loop will be executed at least once.<br>• Exit Do is optional. |

• Do...Loop (test not included):

| Syntax for Do...Loop |
|---|
| Do<br>        {statements that must include how the loop is terminated<br>          and may include Exit Do}<br>Loop |

## • Part C: Branching via If-Then-Else

This statement allows branching according to a true or false result of a condition test. Part E of this section discusses these condition tests. Several types of If-Then-Else are shown below.

• Branching to a single statement

```
If { condition } Then { single statement if condition is true }
```

• Branching to a block of statements

```
If { condition } Then
    { statements if condition is true }
End If
```

• Branching at a fork

```
If { condition } Then
    { statements if condition is true }
Else
    { statements if condition is false }
End If
```

• Branching at a multiple-condition fork

```
If { condition 1 } Then
    { statements if condition 1 is true }
ElseIf { condition 2 } Then
    { statements if condition 1 is false and condition 2 is true }
    Else
    { statements if condition 1 is false and condition 2 is false }
End If
```

• Another type of branching at a multiple-condition fork

```
If { condition 1 } Then
    { statements if condition 1 is true }
ElseIf { condition 2 } Then
    { statements if condition 1 is false and condition 2 is true }
End If
```
Note: Nothing is done if conditions 1 and 2 are both false.

Notes:

- ElseIf doesn't have its own End If.
- There can be many ElseIfs and Elses.

### • **Part D: Branching via Select Case:**

This is branching by comparing the value of a variable to *tags* on several cases. For example, when the variable whichcase = 10, the program branches to Case 10. The following is the syntax.

| Syntax for Select Case |
|---|
| Select Case *whichcase*<br>  Case *tag1*<br>    {statements when *whichcase* = *tag1*, after which control transfers to End Select}<br><br>    .<br><br>  Case *tagn*<br>    {statements when *whichcase* = *tagn*, after which control transfers to End Select}<br>  Case Else<br>    {statements to execute when there is no match}<br>End Select |
| Notes:<br>• *whichcase* is a variable whose value is compared to *tagn*. *whichcase* can be a number (integer or floating point, positive or negative), or it can be a character string.<br>• *tagn* is a constant. |

In Select Case, VBA proceeds down the list of *tags* until it finds a match. After executing the statements in that block, control transfers to End Select, even though other *tags* farther down the list would also match. Hence, the order and the content of the *tags* is very important.

The following code fragment shows the three forms that the *tags* can have. A *tag* can be a number, a range of numbers, or the Is form. The code fragment determines the value of **grade** from the input value of **score**.

| Three Forms of Tags | |
|---|---|
| Code Fragment | Explanation |
| Select Case score<br>  Case 100<br>    grade = " excellent "<br>  Case  90 To 99<br>    grade = " A "<br>  Case  80 To 89<br>    grade = " B "<br>  Case  70 To 79<br>    grade = " C "<br>  Case Is >= 60<br>    grade = " D "<br>  Case Else<br>    grade = " F "<br>End Select | 1) *tag* can be a number (or string).<br><br>2) *tag* can be a range of numbers (or strings), from *lower* to *upper*.<br><br><br><br>3) *tag* can be the Is form and can use any of the *comparison operators* described in part E of this section. |

## • __Part D (cont.):__

The following table shows more examples for *tag*.

| *tag* | Value of whichcase |
|-------|--------------------|
| Case 1 | 1 |
| Case −1 | −1 |
| Case 1, 1.5 | 1 or 1.5  (The comma means *or*) |
| Case 1 To 2 | (from 1 to 2) |
| Case 1 To 2, 4 To 5 | (from 1 to 2) or (from 4 to 5) |
| Case Is = 1 | 1 |
| Case Is = −1 | −1 |
| Case Is = 1, Is = 1.5 | 1 or 1.5 |
| Case Is >= 1, Is <= 2 | Any real number (This is not a good Tag) |
| Case Is < 1 | < 1 including the negative numbers |
| Case "aA a    " | "aA a    "  including spaces between the quotes |

• **Part E: Operators (Comparison, Logical, and Arithmetic)**

In looping and branching, *condition tests* that evaluate as *true* or *false* determine the paths through the code. These *condition tests* use what are called comparison and logical operators. The arithmetic operators include addition and subtraction. This part illustrates the syntax for all of these operators as they apply to numbers. Refer to Lomax[7] for how they apply to character strings.

## Arithmetic Operators

The following is a list of the standard arithmetic operators.

| Operator for... | VBA Code |
|---|---|
| addition | + |
| subtraction and negation | – |
| multiplication | * |
| division | / |
| exponentiation | ^ |

## Comparison Operators

Whereas the result of an arithmetic expression is a number, the result of a comparison expression is *true* or *false*.

| *Comparison Expression* in Words | VBA Syntax | Result of Comparison |
|---|---|---|
| 3 is greater than 2 | 3 > 2 | true |
| 3 is less than 2 | 3 < 2 | false |

The following is a list of all of the VBA comparison operators.

| Operator for the Comparison | VBA Syntax |
|---|---|
| greater than | > |
| less than | < |
| greater than or equal to | > = |
| less than or equal to | < = |
| equal to | = |
| not equal to | <> |

---

[7] Lomax, P., *VB and VBA in a Nutshell* (Sebastopol, CA: O'Reilly & Associates, 1998).

## Logical Operators

Like a comparison expression, the result of a logical expression is true or false. The following table illustrates logical expressions using the logical operator called **And**:

| Logical Expression in Words | VBA Syntax | Result of Comparison |
|---|---|---|
| 3 is greater than 2 And 1 is greater than 0 | 3 > 2 And 1 > 0 | true |
| 3 is greater than 2 And 1 is less than 0 | 3 > 2 And 1 < 0 | false |

The following table illustrates four **VBA** logical operators. In the table, **A** and **B** are logicals, that is, they are true or false.

| VBA Syntax | Result |
|---|---|
| A And B | true if both A and B are true. Otherwise false. |
| A Or B | true if either or both A and B are true. Otherwise false. |
| A Xor B | true if A and B are different. Otherwise false. |
| A Eqv B | true if A and B are the same. Otherwise false. |

Another VBA logical operator is called **Not**. Not-false is true. Not-true is false.

## The Rules for Operator Precedence

• First, the arithmetic is done. Within the arithmetic itself, there is also a precedence.

> • All exponentiation is done first.
> • All multiplication and division are done next and from left to right (e.g., 4 / 2 * 2 = 4).
> • All addition and subtraction are done last, from left to right.

• Second, the comparisons are done.

• Third, the logical expressions are evaluated in the following order:

> Not
> And
> Or
> Xor
> Eqv

When the same operator appears multiple times on the same line, evaluation is from left to right.

## Example Demonstrating Operator Precedence

Consider the following valid VBA statement:

$$b = \text{Not } 2\text{^}2 + 1 > 4 - 2 * 2 / 3 * 3 \text{ And } 1 < 0 \text{ Or } 3 > 2 \qquad \text{Equation 1}$$

This statement is from bad programming, but it is valid. To evaluate it, the obvious first step is to do the arithmetic. The statement becomes as follows:

$$b = \text{Not } 5 > 0 \text{ And } 1 < 0 \text{ Or } 3 > 2$$

This statement starts with **Not 5**, which by itself doesn't make sense. However, by using parentheses for grouping, $\quad b = \text{Not } (5 > 0) \text{ And } (1 < 0) \text{ Or } (3 > 2)$

$$b = \textit{false} \text{ And } \textit{false} \text{ Or } \textit{true}$$

Now comes a dilemma. Evaluating **b** from left to right, this results.

$$b = \textit{false} \text{ Or } \textit{true} = \textit{true}$$

Evaluating **b** from right to left, this results.

$$b = \textit{false} \text{ And } \textit{true} = \textit{false}$$

There must be a rule to tell which one. There is! **And** is evaluated before **Or**. Hence, the following is accurate.

$$b = \textit{false} \text{ Or } \textit{true} = \text{true}$$

Is that what the programmer wants?

Tip: Expressions within a parentheses are evaluated first.

So, rather than memorize rules, the programmer can and should use parentheses to organize the conditions to get the right answer.

For example, **b = True** if the parentheses in **Equation 1** are as follows.

$$b = \left(\text{Not}\left((2\text{^}2 + 1) > (4 - 2 * (2 / 3) * 3)\right) \text{ And } 1 < 0\right) \text{ Or } (3 > 2)$$

**b = False** if the parentheses in **Equation 1** are as follows.

$$b = \left(\text{Not}\left((2\text{^}2 + 1) > (4 - 2 * (2 / 3) * 3)\right)\right) \text{ And } (1 < 0 \text{ Or } 3 > 2)$$

Note: The following VBA statements are valid.

```
b = True
Cells(1, 1) = b
```

The logical variable **True** is printed in cell(1, 1).

## Section 8.7: The Call Statement

This section shows how a program can be modularized. A main program calls a subprogram to do a task. After the task, control returns to the main. The following table illustrates program-flow when using the Call statement.

| | | |
|---|---|---|
| Start<br>Step 1<br>Step 2<br>Step 7<br>Step 8 | Sub AAA()<br>  a2 = 4<br>  Call BBB(b2, a2)<br>  Cells(2, 1) = b2<br>End Sub | At Step 2, the program transfers to the subprogram called BBB. Enclosed within parentheses are the variables a2 and b2. In Sub BBB, a2 is printed and b2 is set to a value. Then the program returns to Sub AAA where b2 is printed and the program ends. |
| Step 3<br>Step 4<br>Step 5<br>Step 6 | Sub BBB(b2, a2)<br>  Cells(1, 1) = a2<br>  b2 = 5<br>End Sub | |

### Notes
- The argument list contains all of the variables that are transferred to and from.
- There is no limit on the number of Call statements or subprograms.
- Subprograms can call other subprograms.
- Except for the argument list, values of the variables in a subprogram must be reset or recomputed each time the subprogram is entered.
- The argument list may be empty.
- The calling program and the called program should be in the same code window.
- Only the programs with empty parentheses are listed in the Macros window.

The following table contains a listing of an easy-to-follow program. It starts with a main program called **zero**. There are two subprograms called **one** and **two**. These three parts pass around a variable named A1. Starting with A1 = 0, each program increases the value of A1 until the final value is printed out in **zero** and the whole program ends. Read the code and satisfy yourself that the final value is A1 = 9.

| Program Call_1 | |
| --- | --- |
| Sub zero()<br>  A1 = 0<br>  Call one(A1)<br>  Cells(1, 1) = A1<br>End Sub<br><br>Sub one(A1)<br>  A1 = A1 + 3<br>  Call two(A1)<br>  A1 = A1 + 3<br>End Sub<br><br>Sub two(A1)<br>  A1 = A1 + 3<br>End Sub | Note that the three parts should be in the same code window. |

The following program produces identical results to the above program. It shows that the variable A1 is really an address in memory. In **Sub one**, this address is called B1. In **Sub two**, it is called C1.

| Program Call_2 |
| --- |
| Sub zero()<br>  A1 = 0<br>  Call one(A1)<br>  Cells(1, 1) = A1<br>End Sub<br><br>Sub one(B1)<br>  B1 = B1 + 3<br>  Call two(B1)<br>  B1 = B1 + 3<br>End Sub<br><br>Sub two(C1)<br>  C1 = C1 + 3<br>End Sub |

Program Call_2 uses the single variable A1. The following is a similar program that declares A1 to be a matrix that can contain fifty numbers. The program demonstrates several other features about the Dim statement.

<table>
<tr><td colspan="2" align="center">Program Call_3</td></tr>
<tr>
<td>

```
Sub zero()
Dim A1(50)
  xx = Array(-8, 3, 1)
  A1(4) = 0
  Call one(A1)
  Cells(1, 1) = A1(4)
  Cells(2, 1) = xx(0)
End Sub

Sub one(B1)
  B1(4) = B1(4) + 3
  Call two(B1)
  B1(4) = B1(4) + 3
End Sub

Sub two(C1)
Dim D1(10)
  D1(5) = 3
  C1(4) = C1(4) + D1(5)
End Sub
```

</td>
<td>

• Before a matrix can be used, it must be declared in a Dim Statement. This declaration appears only in the program (or subprogram) wherein the matrix first appears.

• Matrix D1 that is declared in Sub two, is only valid in Sub two.

• The final value for A1(4) is 9.

• Note that xx(0) is printed out. Its value = −8. Though it acts like a matrix, xx comes from an Array function. The thing to know is that the output of an Array function does not appear in a Dim statement.

• See section 8.3 for more on the Dim statement.

</td>
</tr>
</table>

The following table shows additional VBA syntax.

---

• **Exit Sub:** This optional statement causes execution to return to the calling program. Execution returns to the statement right after the **Call** statement. If used in the main program, it causes execution to stop.

• **End:** This optional statement causes execution to stop. It can be used anywhere.

---

# Section 8.8: The GoSub Statement

This section shows another way a program can be modularized. The following table contains a demonstration.

| GoSub Demonstration Program | |
| --- | --- |
| Code | Comments |
| Sub main()<br><br>   a = 10<br>   GoSub one<br>   GoSub two<br>   End<br><br>one:<br>   Cells(1, 1) = a<br>   Return<br><br>two:<br>   b = 20<br>   Cells(2, 1) = b<br>   Return<br><br>End Sub | • The statement, GoSub one, sends the program to the statement labeled *one*. There, a is printed. Then the Return statement sends the program back to the statement after GoSub one.<br><br>• GoSub two sends the program to the statement labeled *two*. There, b is set to 20 and is printed. Then the Return statement sends the program back to the statement after GoSub two.<br><br>• End terminates the program. This prevents looping. |
| The syntax for a statement label<br>It must be some unique name followed by a *colon*. It must be the first entry on a line. (VBA will reposition it to column 1.) There's no limit on the number of labels. | |
| Note: The GoSub statement may also be used in a subprogram or a function. | |

# Section 8.9: The GoTo Statement
## with Sort and Interpolate

This section defines the GoTo statement. Simply stated, GoTo acts like GoSub without the Return. The following table contains a demonstration program.

| GoTo Demonstration Program | |
|---|---|
| **Code** | **Comments** |
| Sub main()<br>    GoTo one<br><br>two:<br>    Cells(1, 1) = a<br>    End<br><br>one:<br>    a = 10<br>    GoTo two<br>End Sub | • GoTo one sends the program to the statement labeled one. There, a is set to 10. Then the GoTo two statement sends the program to the statement labeled two.<br><br>• At two, a is printed.<br><br>• End terminates the program. This prevents looping. |
| The Syntax for a Statement Label<br>It must be some unique name followed by a *colon*. It must be the first entry on a line. (VBA will reposition it to column 1.) There's no limit on the number of labels. | |
| Note: The GoTo statement may also be used in a subprogram or function. | |

The following example uses the GoTo statement.

**Example GoTo:** Take a list of paired numbers, Xin(i) and Yin(i). Sort Xin(i) in ascending order.

| Unsorted | | Sorted | |
|---|---|---|---|
| Xin(i) | Yin(i) | Xin(i) | Yin(i) |
| 9 | 3 | -9 | -3 |
| -9 | -3 | -7 | -1 |
| 7 | 1 | -5 | 1 |
| -7 | -1 | -3 | 3 |
| 5 | 1 | -1 | 5 |
| -5 | 1 | 1 | 5 |
| 3 | 3 | 3 | 3 |
| -3 | 3 | 5 | 1 |
| 1 | 5 | 7 | 1 |
| -1 | 5 | 9 | 3 |

Then by interpolating in this sorted list, find the **y** that corresponds to a value input for **x**. For this example, use x = −4.

The following table lists the program for **Example GoTo**. The **GoTo** statement is used in the interpolation segment.

| Program for Example GoTo | |
| --- | --- |
| Code | Comments |
| `Sub main()`<br>`  Dim Xin(100), Yin(100)`<br>`  n = Cells(1, 1):    x = Cells(1, 6)`<br>`  For i = 1 To n`<br>`    Xin(i) = Cells(i, 2):  Yin(i) = Cells(i, 3)`<br>`  Next i`<br>`  For L = n To 2 Step -1`<br>`    For i = 1 To L - 1`<br>`      If Xin(i) > Xin(L) Then`<br>`        temp = Xin(i)`<br>`        Xin(i) = Xin(L)`<br>`        Xin(L) = temp`<br>`        temp = Yin(i)`<br>`        Yin(i) = Yin(L)`<br>`        Yin(L) = temp`<br>`      End If`<br>`    Next i`<br>`  Next L`<br>`  For i = 1 To n`<br>`    Cells(i, 4) = Xin(i): Cells(i, 5) = Yin(i)`<br>`  Next i`<br>`  If (x < Xin(1)) Or (x > Xin(n)) Then`<br>`    Cells(1,7) = " x out of bounds": End`<br>`  End If`<br>`  For i = 2 To n`<br>`    If x <= Xin(i) Then`<br>`      m = i - 1`<br>`      GoTo st200`<br>`    End If`<br>`  Next i`<br>`  st200:`<br>`  ydel = (Yin(m + 1) - Yin(m)) * (x - Xin(m)) / (Xin(m + 1) - Xin(m))`<br>`  y = Yin(m) + ydel`<br>`  Cells(1, 7) = y`<br>`End Sub` | • n is the number of points, and x is the input.<br>• The unsorted pair is $X_{in}$ and $Y_{in}$.<br><br>• Sort<br><br>  • Sort into ascending order.<br>  • Outer loop runs from bottom up.<br>  • Inner loop runs from top down.<br>  • First pass puts largest number on the bottom.<br>  • Second pass puts second-largest number second last.<br>  • This pattern continues.<br><br>• Print the sorted pair.<br><br>• Check the bounds.<br><br>• Interpolate<br><br>  Beginning at the smallest number,<br>  when x is between Xin(m)  and Xin(m+1),<br>  the GoTo st200 statement transfers<br>  execution to the statement labeled st200.<br><br>• Linear interpolation is used.<br><br>• y is the output that corresponds to the input x. |

The following is the scatter chart of the sorted input list and the result.

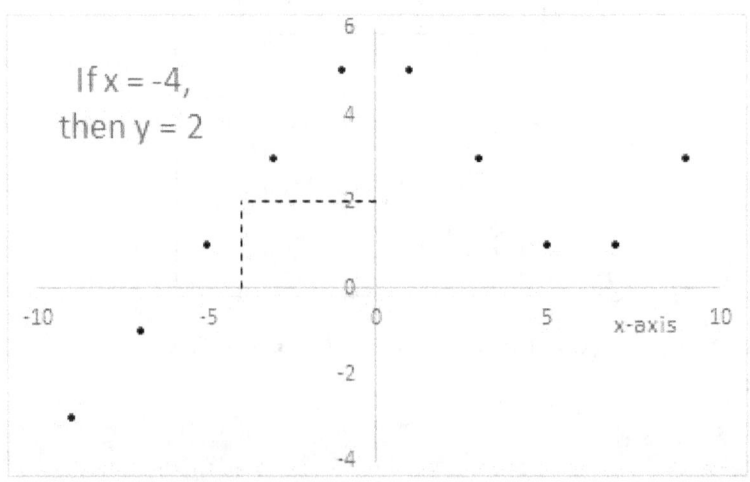

# Section 8.10: Debugging

Debugging is fun. Point-and-click menus and windows make it easy. Execution can be stopped in several ways and at any line of code. Values of variables can be inspected in several ways. Execution can then be continued until the next stopping point. A simple program shows all of this.

The following table shows the listing and description of the program that will be analyzed by debugging. The program merely passes a variable around a loop between a main program and two subprograms. Type this program into a code window.

| The Program Used to Demonstrate Debugging | |
|---|---|
| Code | Description |
| ```
Sub main()
  For w = 1 To 5
    Call one(w, zDum)
    Cells(w, 1) = zDum
  Next w
End Sub

Sub one(wDum, z)
  x = wDum
  Call two(x, yDum)
  z = yDum
End Sub

Sub two(xDum, y)
  y = xDum
End Sub
``` | • In Sub main, the For statement starts with w = 1.<br>• Call Sub one with w in the argument list.<br>• Sub one<br>  • wDum is equivalenced to w in the argument list.<br>  • x = wDum.<br>  • x is passed to Sub two.<br>• Sub two<br>  • xDum is equivalenced to x.<br>  • y = xDum.<br>  • y is passed back to Sub one.<br>• Sub one<br>  • yDum is equivalenced to y.<br>  • z = yDum.<br>  • z is passed back to Sub main.<br>• Sub main<br>  • zDum is equivalenced to z.<br>  • zDum is printed out.<br>• The Next statement then sets w = 2, and execution continues on the second pass. |

The first step in debugging is to open the **Watches** window.

Click **Editor**, and then in the View menu, click **Watch Window**.

**Watches** will show the values of selected variables as the program executes. For this demonstration, all of the variables in the above program will be put into **Watches**. The following are the instructions on how to do this.
```

- Open the code window that contains the above program.
- Put the cursor next to a variable (e.g., w).
- Then in the Debug menu, click Add Watch. The Add Watch window pops up, and the variable w is already entered. (Options in the Add Watch window will be discussed later.)
- Click OK. The Add Watch window closes. The variable w now appears in Watches and is out of context since the program is not yet executing.
- Put the cursor next to another variable (e.g., zDum). Repeat the process until Watches looks like the following table.

| The Watches Window before Execution | | | |
|---|---|---|---|
| Expression | Value | Type | Context |
| w | Out of Context | Variant/Empty | ThisWorkbook.main |
| zDum | Out of Context | Variant/Empty | ThisWorkbook.main |
| z | Out of Context | Variant/Empty | ThisWorkbook.one |
| wDum | Out of Context | Variant/Empty | ThisWorkbook.one |
| x | Out of Context | Variant/Empty | ThisWorkbook.one |
| yDum | Out of Context | Variant/Empty | ThisWorkbook.one |
| xDum | Out of Context | Variant/Empty | ThisWorkbook.two |
| y | Out of Context | Variant/Empty | ThisWorkbook.two |
| Note: The Watches window is not saved when the program file is closed. | | | |

Variables can be added to the Watches window at any time during debug. Now to begin debugging.

Open the code window and put the cursor anywhere in Sub main. Select the Debug menu. The focus is now on four commands: Step Into, Step Over, Step Out, and Run To Cursor. We will start with the Step Into command.

- Step Into: In the Debug menu, click Step Into. This Step Into command puts the arrow on line 1 of the code. At the next Step Into command, line 1 will execute, and the cursor will point to line 2. The following two tables show each successive Step Into command, and the values of the variables after each step.

This table shows the code line after each **Step Into** command.

| The Code Window | |
|---|---|
| Step Into *Command Numbers* | Code Lines |
| 1 | Sub main() |
| 2 |   For w = 1 To 5 |
| 3 and 14 |     Call one(w, zDum) |
| 12 |     Cells(w, 1) = zDum |
| 13 |   Next w |
| | End Sub |
| 4 | Sub one(wDum, z) |
| 5 |   x = wDum |
| 6 |   Call two(x, yDum) |
| 10 |   z = yDum |
| 11 | End Sub |
| 7 | Sub two(xDum, y) |
| 8 |   y = xDum |
| 9 | End Sub |

This table lists the value of the variables after each **Step Into** command.

| The Watches Window | | | | | | | | | | | | | | | |
|---|---|---|---|---|---|---|---|---|---|---|---|---|---|---|---|
| | | The Value *after* Each Step Into Command | | | | | | | | | | | | | |
| | Start | 1 | 2 | 3 | 4 | 5 | 6 | 7 | 8 | 9 | 10 | 11 | 12 | 13 | 14 |
| w | O | E | E | 1 | 1 | 1 | 1 | 1 | 1 | 1 | 1 | 1 | 1 | 1 | 2 |
| zDum | O | E | E | E | E | E | E | E | E | E | E | 1 | 1 | 1 | 1 |
| z | O | O | O | O | E | E | E | E | E | E | E | 1 | O | O | O |
| wDum | O | O | O | O | 1 | 1 | 1 | 1 | 1 | 1 | 1 | 1 | O | O | O |
| x | O | O | O | O | E | E | 1 | 1 | 1 | 1 | 1 | 1 | O | O | O |
| yDum | O | O | O | O | E | E | E | E | E | 1 | 1 | 1 | O | O | O |
| xDum | O | O | O | O | O | O | O | 1 | 1 | 1 | O | O | O | O | O |
| y | O | O | O | O | O | O | O | E | E | 1 | O | O | O | O | O |
| *Notes:* • O is the legend for Out of Context, that is, in a part of the program that is not active. <br> • E is the legend for Empty. | | | | | | | | | | | | | | | |

The tables show that after the third **Step Into** command, line 2 has executed and w = 1. Continue analyzing the above table, or go to the **Run** menu and **Reset**.

## The Other Options in the Debug Menu

• <u>Step Over:</u> When the **Debug** arrow is pointing to a **Call** statement (or a **Function** statement), the **Step Over** command causes the debugger to execute the called program and stop on the statement after the **Call**.

• <u>Step Out:</u> When the **Debug** arrow is pointing to any statement in a subprogram or function, the **Step Out** command causes the debugger to execute the rest of the subprogram and stop on the statement after the **Call**. Note that when the **Debug** arrow is pointing to any statement in the main program, the **Step Out** command causes the rest of the program to be completed and stop.

• <u>Run To Cursor:</u> Put the cursor at the beginning or end of any statement. The **Run To Cursor** command causes the debugger to execute all statements up to the cursor and stop at the cursor.

## Viewing Variables with the Cursor

This is another way to see the values of variables. Suppose the **Debug** arrow is pointing to the following statement: **Call one(w, zDum)**. Hold the mouse pointer over the variable w. After a slight delay, a pop-up will show the value of w. In fact, holding the mouse pointer over any variable for a moment will produce a pop-up showing the value, if it's available.

## Debugging by Run To Breakpoint

Another way of debugging is to set breakpoints. The following table describes how to set breakpoints.

| Code Window Illustrating Breakpoints | |
|---|---|
| Code | Description |
| Sub main()<br>　For w = 1 To 5<br>　　Call one(w, zDum)<br>　　Cells(w, 1) = zDum<br>　Next w<br>End Sub<br>Sub one(wDum, z)<br>•　x = wDum<br>　Call two(x, yDum)<br>　z = yDum<br>End Sub<br>Sub two(xDum, y)<br>•　y = xDum<br>End Sub | • A breakpoint is set by clicking in the gray area in front of a line of code.<br><br>• Two breakpoints (bullets) are shown.<br><br>• Like a toggle switch, another click will turn it off.<br><br>• The program need not be in the debug mode when setting a breakpoint. |

With the breakpoints as shown, Run the program. Execution will stop at the statement x = wDum. Values of the variables can be inspected either in Watches or with the mouse pointer. Then, Run Menu → Continue causes execution to the next breakpoint. A debugging session can include using the options from the Debug menu together with run-to-breakpoint. The menus always show the options that are available. When debugging is finished, Run Menu → Reset.

### Options in the Add Watch Window

Besides variables, the Watches window can include arithmetic, comparative, and logical expressions (See section 8.6). The Add Watch window allows one of the following options to be selected for each expression:

- Watch Expression (This is the option that has been used so far in this section.)
- Break When Value Is True
- Break When Value Changes

The following are examples of expressions that can be put into the Watches window. Never mind if they don't make sense.

| | |
|---|---|
| w | a variable |
| x + wDum - y * z | an arithmetic expression |
| y > 3 | a comparison expression |
| w > 3  And zDum > 3 | a logical expression |

### The Immediate Window

This window can display program output via special VBA debug statements that are placed in the code. This window will not be discussed in this book.

### Final Note

Back to the oxymoron that *debugging is fun*. When the code malfunctions, VBA automatically stops in the debug mode, points an arrow at the difficult line of code, and allows values to be inspected by hovering the cursor above any variable. And of course, the debug mode can be used as a learning-aide even when there is no malfunction.

## Section 8.11: List of Elements in the VBA Language

| Statements | Common VBA Functions (Section 8.5) |
|---|---|
| | Array, also section 8.2 and 8.3 |
| Call, section 8.7 | Abs |
| Dim, section 8.3 | Atn |
| Do … Loop, section 8.6 | Cos |
| Do Until … Loop, section 8.6 | Exp |
| Do … Loop Until, section 8.6 | Fix |
| Do While … Loop, section 8.6 | Int |
| Do … Loop While, section 8.6 | Log |
| End, section 8.7 | Round |
| End Function, section 8.5 | Sgn |
| End If, section 8.6 | Sin |
| End Select, section 8.6 | Sqr |
| End Sub, the last statement in all programs | Tan |
| Exit Do, section 8.6 | **Common Excel Functions (Section 8.5)** |
| Exit For, section 8.6 | Asin |
| Exit Function, section 8.5 | Acos |
| Exit Sub, section 8.7 | Atan2 |
| For … To … Next, section 8.6 | Log10 |
| Function, Excel, section 8.5 | Log |
| Function, VBA, section 8.5 | Max |
| Function, User-defined, section 8.5 | Min |
| GoSub … Return, section 8.8 | Minverse, also section 8.4 |
| GoTo, section 8.9 | MMult, also section 8.4 |
| If … Then, section 8.6 | MDeterm, also section 8.4 |
| If … Then … End If, section 8.6 | Transpose |
| If … Then … Else … End If, section 8.6 | Round |
| If … Then … ElseIf … Then … Else … End If, section 8.6 | RoundUp |
| Option Base, section 8.3 | RoundDown |
| Option Explicit, section 8.3 | **Complex Excel Functions (Section 8.5)** |
| ReDim, section 8.4 | Complex |
| Select Case, section 8.6 | ImReal |
| Sub, the first statement in all programs | Imaginary |
| | ImAbs |
| | ImArgument |
| | ImDiv |
| | ImProduct |
| | ImSum |
| | ImSub |
| | ImPower |
| | ImExp |

| Prefixes on the Cells(row,column) Statement in Section 8.1 |
|---|
| Sheets( $j$ ).Cells(row, column)<br>Worksheets("sheet$j$").Cells(row, column) |

| Operators Used in Arithmetic and in Conditional Tests (Comparison/Logical) |
|---|
| Section 8.6 |

# Appendices

# Appendix A: Derivation of the Quadratic Formula

The following is a derivation of the famous Quadratic formula that finds the zeros of the equation:

$$x^2 + bx + c = 0 \qquad \text{where b and c are real constants.}$$

$$x^2 + bx = -c$$

$$x^2 + bx + \frac{b^2}{4} = -c + \frac{b^2}{4}$$

$$\left(x + \frac{b}{2}\right)^2 = \frac{b^2 - 4c}{4}$$

$$\left(x + \frac{b}{2}\right) = \pm\frac{\sqrt{b^2 - 4c}}{2}$$

Finally, $\qquad x_1 = \dfrac{-b + \sqrt{b^2 - 4c}}{2} \quad \text{and} \quad x_2 = \dfrac{-b - \sqrt{b^2 - 4c}}{2}$

In factored form, $\qquad (x - x_1)(x - x_2) = 0$

# Appendix B: A Program to Find All the Zeros of a Polynomial

We will start with a discussion of the Eigenvalue problem, which is to find the values of $\lambda$ that solve the following algebraic equation.

$$A\overline{x} = \overline{x}\lambda$$

- A is a square matrix.
- $\overline{x}$ is a vector of the unknown variables.
- $\lambda$ is a scalar called an eigenvalue.

Wilkinson[8] is a thorough reference for this problem. Rearranging the above equation,

$$[A - \lambda I]\overline{x} = 0, \text{ where } I \text{ is identity matrix.}$$

This is a homogeneous equation. We know from chapter 3 that it has a non-zero solution only if

$$|A - \lambda I| = 0.$$

From chapter 5, we know that $|A - \lambda I| = 0$ is polynomial equation.

$$\lambda^n + k_{n-1}\lambda^{n-1} + k_{n-2}\lambda^{n-2} + \bullet\bullet\bullet + k_1\lambda + k_0 = 0$$

If A is an upper-triangular matrix, then $A = \begin{bmatrix} a_{11} & a_{12} & a_{13} & \bullet & a_{1n} \\ 0 & a_{22} & a_{23} & \bullet & a_{2n} \\ 0 & 0 & a_{33} & \bullet & a_{3n} \\ \bullet & \bullet & \bullet & \bullet & \bullet \\ 0 & 0 & 0 & \bullet & a_{nn} \end{bmatrix}$.

Hence, $|A - \lambda I| = \begin{vmatrix} (a_{11} - \lambda) & a_{12} & a_{13} & \bullet & a_{1n} \\ 0 & (a_{22} - \lambda) & a_{23} & \bullet & a_{2n} \\ 0 & 0 & (a_{33} - \lambda) & \bullet & a_{3n} \\ \bullet & \bullet & \bullet & \bullet & \bullet \\ 0 & 0 & 0 & \bullet & (a_{nn} - \lambda) \end{vmatrix} = 0.$

From chapter 5, this can be written

$$|A - \lambda I| = (a_{11} - \lambda)(a_{22} - \lambda)(a_{33} - \lambda) \bullet\bullet\bullet (a_{nn} - \lambda) = 0$$

Francis[9] developed the QR method to turn a general matrix into upper-triangular form. Therefore, if we change a polynomial into a matrix and apply the QR method, we get the zeros of the polynomial.

---

[8] Wilkinson, J., *The Algebraic Eigenvalue Problem* (Oxford, UK: Oxford University Press, 1965).
[9] Francis, J., *The QR Transformation, Parts I and II* (The Computer Journal, Vol. 4, 1961, pages 265-271).

In chapter 5, we showed how to change a polynomial into a matrix. In this appendix, we will present the QR method. In the following discussion, zeros of a polynomial are called eigenvalues of a matrix.

Rutishauser[10] discovered how to iteratively transform a general matrix into upper-triangular form while preserving its eigenvalues. This process is known as the LR method.

Starting with a matrix $A_o$, the first step is to factor it into a lower-triangular matrix $L_o$ and an upper-triangular matrix $R_o$.

$$A_o = L_o * R_o .$$

I discuss later how to compute $L_o$ and $R_o$. Because they have the same $\lambda$'s, $A_o$ and $(L_o * R_o)$ are similar. From the definition of similarity, this means that for *any* matrix T,

$$T * A_o = (L_o * R_o) * T .$$

Using $L_o$ as the T matrix, $\qquad\qquad L_o * A_o = (L_o * R_o) * L_o .$

Then, $\qquad\qquad A_o = L_o^{-1} * (L_o * R_o) * L_o = R_o * L_o .$

This new matrix $(R_o * L_o)$ comes from multiplying the factors of $A_o$ in reverse. It has the same eigenvalues as the original $A_o$ matrix. The LR method now uses this new matrix as the starting point for the next iteration. This is shown in the following figure.

Rutishauser proved that if you continue this process and there are no numerical difficulties, the following will happen. The matrix L will approach the identity matrix. When this happens, A equals R and is therefore upper-triangular. Its eigenvalues are the same as the original A matrix and are on its diagonal.

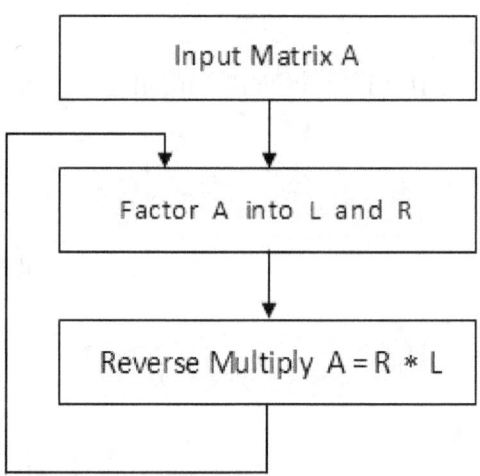

Appendix C shows a program that implements the LR method. Let's illustrate it with the following example.

Example LR1: Use the LR program to find the zeros of the following polynomial.

$$x^3 + 2.5x^2 - 11.5x + 5$$

In matrix form, $\qquad\qquad A = \begin{bmatrix} 0 & 1 & 0 \\ 0 & 0 & 1 \\ -5 & 11.5 & -2.5 \end{bmatrix}$

The following table shows the results.

---

[10] Rutishauser, H., *Solution of Eigenvalue Problems with the LR Transformation* (Nat. Bur. of Standards Appl. Math Series, Vol 49, 1958, pages 47-81).

| | Example LR1 | | | (*Note:* $A_k = L_k * R_k$ and $A_{k+1} = R_k * L_k$) | | | | | |
|---|---|---|---|---|---|---|---|---|---|
| Iteration (k) | The A Matrices at Start of Each Iteration | | | The L Matrices (Lower Triangular) | | | | The R Matrices (Upper Triangular) | |
| 0 | 0 | 1 | 0 | 1 | 0 | 0 | 0 | 1 | 0 |
| | 0 | 0 | 1 | 0 | 1 | 0 | 0 | 0 | 1 |
| | -5 | 11.5 | -2.5 | -5000 | 5.0E+6 | 1 | 0 | 0 | -5.0E6 |
| 1 | 0 | 1 | 0 | 1 | 0 | 0 | 0 | 1 | 0 |
| | -5000 | 5.0E+06 | 1 | -5.0E+6 | 1 | 0 | 0 | 1.0E7 | 1 |
| | 2.5E+10 | -2.5E+13 | -5.0E+06 | 2.5E+13 | -5.0E6 | 1 | 0 | 0 | 0 |
| 2 | -5.0E06 | 1 | 0 | 1 | 0 | 0 | -5.0E6 | 1 | 0 |
| | -2.5E+13 | 5.0E+06 | 1 | 5.0E6 | 1 | 0 | 0 | 0 | 1 |
| | -1.3E+10 | 2508.63 | 0 | 2508.63 | 0 | 1 | 0 | 0 | 0 |
| 3 | -2.5 | 1 | 0 | 1 | 0 | 0 | -2.5 | 1 | 0 |
| | 11.51 | 0 | 1 | -4.61 | 1 | 0 | 0 | 4.61 | 1 |
| | -5.03 | 0 | 0 | 2.02 | -0.44 | 1 | 0 | 0 | 0.44 |
| 4 | -7.11 | 1 | 0 | 1 | 0 | 0 | -7.11 | 1 | 0 |
| | -19.24 | 4.17 | 1 | 2.71 | 1 | 0 | 0 | 1.47 | 1 |
| | 0.88 | -0.19 | 0.44 | -0.12 | -0.05 | 1 | 0 | 0 | 0.48 |
| 5 | -4.4 | 1 | 0 | 1 | 0 | 0 | -4.4 | 1 | 0 |
| | 3.85 | 1.42 | 1 | -0.87 | 1 | 0 | 0 | 2.29 | 1 |
| | -0.06 | -0.02 | 0.48 | 0.01 | -0.02 | 1 | 0 | 0 | 0.5 |
| 6 | -5.27 | 1 | 0 | 1 | 0 | 0 | -5.27 | 1 | 0 |
| | -1.99 | 2.28 | 1 | 0.38 | 1 | 0 | 0 | 1.9 | 1 |
| | 0.01 | -0.01 | 0.5 | 0 | 0 | 1 | 0 | 0 | 0.5 |
| 7 | -4.9 | 1 | 0 | 1 | 0 | 0 | -4.9 | 1 | 0 |
| | 0.72 | 1.9 | 1 | -0.15 | 1 | 0 | 0 | 2.04 | 1 |
| | 0 | 0 | 0.5 | 0 | 0 | 1 | 0 | 0 | 0.5 |
| 8 | -5.04 | 1 | 0 | 1 | 0 | 0 | -5.04 | 1 | 0 |
| | -0.3 | 2.04 | 1 | 0.06 | 1 | 0 | 0 | 1.98 | 1 |
| | 0 | 0 | 0.5 | 0 | 0 | 1 | 0 | 0 | 0.5 |
| 9 | -4.98 | 1 | 0 | 1 | 0 | 0 | -4.98 | 1 | 0 |
| | 0.12 | 1.98 | 1 | -0.02 | 1 | 0 | 0 | 2.01 | 1 |
| | 0 | 0 | 0.5 | 0 | 0 | 1 | 0 | 0 | 0.5 |
| 10 | -5.01 | 1 | 0 | 1 | 0 | 0 | -5.01 | 1 | 0 |
| | -0.05 | 2.01 | 1 | 0.01 | 1 | 0 | 0 | 2 | 1 |
| | 0 | 0 | 0.5 | 0 | 0 | 1 | 0 | 0 | 0.5 |
| 11 | -5 | 1 | 0 | 1 | 0 | 0 | -5 | 1 | 0 |
| | 0.02 | 2 | 1 | 0 | 1 | 0 | 0 | 2 | 1 |
| | 0 | 0 | 0.5 | 0 | 0 | 1 | 0 | 0 | 0.5 |
| 12 | -5 | 1 | 0 | 1 | 0 | 0 | -5 | 1 | 0 |
| | -0.01 | 2 | 1 | 0 | 1 | 0 | 0 | 2 | 1 |
| | 0 | 0 | 0.5 | 0 | 0 | 1 | 0 | 0 | 0.5 |
| 13 | -5 | 1 | 0 | | | | | | |
| | 0 | 2 | 1 | | | | | | |
| | 0 | 0 | 0.5 | | | | | | |

By iteration 13, A has been triangularized and the $\lambda$'s are on its diagonal. They are: $\lambda = 0.5, 2$, and $-5$. Note that the zeros in most of the above matrices are actually small numbers, for example, $1E-6$.

**Example LR2:** Use the **LR** program to find the zeros of $x^3 - 22x^2 + 144x - 288$. In matrix form,

$$A = \begin{bmatrix} 0 & 1 & 0 \\ 0 & 0 & 1 \\ 288 & -144 & 22 \end{bmatrix}$$

The following table shows the results.

| Example LR2 | | | | (Note: $A_k = L_k * R_k$ and $A_{k+1} = R_k * L_k$) | | | | | | |
|---|---|---|---|---|---|---|---|---|---|---|
| Iteration (k) | The A Matrices at Start of Each Iteration | | | | The L Matrices (Lower Triangular) | | | | The R Matrices (Upper Triangular) | |
| 0 | 0 | 1 | 0 | | 1 | 0 | 0 | | 0 | 1 | 0 |
| | 0 | 0 | 1 | | 0 | 1 | 0 | | 0 | 0 | 1 |
| | 288 | -144 | 22 | | 288000 | -2.9E8 | 1 | | 0 | 0 | 2.88E8 |
| 1 | 0 | 1 | 0 | | 1 | 0 | 0 | | 0 | 1 | 0 |
| | 288000 | -2.9E+08 | 1 | | 2.88E+8 | 1 | 0 | | 0 | -5.8E8 | 1 |
| | 8.3E+13 | -8.3E+16 | 2.88E+8 | | 8.3E16 | 2.88E8 | 1 | | 0 | 0 | 0 |
| 2 | 2.88E+08 | 1 | 0 | | 1 | 0 | 0 | | 2.88E8 | 1 | 0 |
| | -8.3E+16 | -2.9E+08 | 1 | | -2.9E8 | 1 | 0 | | 0 | 0 | 1 |
| | -4.1E+13 | -144096 | 0 | | -144096 | 0 | 1 | | 0 | 0 | 0 |
| · · · | | | | | | | | | | |

The **LR** method fails. This is due to division by small numbers, and the **LR** method implicitly involves matrix inversion.

The **QR** method by Francis avoids division-by-zero by replacing L with an orthogonal matrix. The inverse of an orthogonal matrix is its transpose. Francis calls this matrix the Q matrix: $A = Q * R$, where R is still an upper triangular. This kind of factorization is discussed in a work by Hager[11]. The **QR** method is otherwise exactly like the **LR**. A program that implements the **QR** is shown in appendix D.

Let's apply the **QR** program to the polynomial where the **LR** method failed. The results are shown in the following table.

---

[11] Hager, W., *Applied Numerical Linear Algebra* (Upper Saddle River, New Jersey: Prentice-Hall, 1988).

| Example QR1 | | | (*Note:* A$_k$ = Q$_k$ * R$_k$ and A$_{k+1}$ = R$_k$ * Q$_k$) | | | | | | |
|---|---|---|---|---|---|---|---|---|---|
| Iteration (k) | The A Matrices at Start of Each Iteration | | | | The Q Matrices (Orthogonal) | | | The R Matrices (Upper Triangular) | |
| 0 | 0 | 1 | 0 | | 0 | -1 | 0 | -288 | 144 | -22 |
| | 0 | 0 | 1 | | 0 | 0 | 1 | 0 | -1 | 0 |
| | 288 | -144 | 22 | | -1 | 0 | 0 | 0 | 0 | 1 |
| 1 | 22 | 288 | 144 | | -1 | -0.05 | 0 | -22.02 | -287.7 | -143.85 |
| | 0 | 0 | -1 | | 0 | 0 | -1 | 0 | -13.08 | -6.54 |
| | -1 | 0 | 0 | | 0.05 | -1 | 0 | 0 | 0 | 1 |
| 2 | 15.47 | 144.7 | 287.7 | | -1 | -0.02 | 0 | -15.47 | -144.55 | -287.4 |
| | -0.3 | 6.53 | 13.08 | | 0.02 | -0.99 | 0.15 | 0 | -9.42 | -18.51 |
| | 0.05 | -1 | 0 | | 0 | 0.15 | 0.99 | 0 | 0 | 1.98 |
| 3 | 13.54 | 99.72 | -305.95 | | -1 | -0.01 | 0 | -13.54 | -99.65 | 305.75 |
| | -0.13 | 6.51 | -19.72 | | 0.01 | -1 | -0.05 | 0 | -7.45 | 22.47 |
| | -0.01 | 0.3 | 1.95 | | 0 | -0.05 | 1 | 0 | 0 | 2.86 |
| 4 | 12.74 | 85.65 | 310 | | -1 | 0 | 0 | -12.74 | -85.62 | -309.89 |
| | -0.06 | 6.41 | 22.79 | | 0 | -1 | 0.02 | 0 | -6.81 | -24.18 |
| | 0 | -0.13 | 2.85 | | 0 | 0.02 | 1 | 0 | 0 | 3.32 |
| 5 | 12.37 | 79.32 | -311.57 | | -1 | 0 | 0 | -12.37 | -79.31 | 311.52 |
| | -0.03 | 6.32 | -24.32 | | 0 | -1 | -0.01 | 0 | -6.51 | 25.03 |
| | 0 | 0.07 | 3.32 | | 0 | -0.01 | 1 | 0 | 0 | 3.58 |
| 6 | 12.18 | 75.99 | 312.39 | | -1 | 0 | 0 | -12.18 | -75.98 | -312.32 |
| | -0.01 | 6.24 | 25.09 | | 0 | -1 | 0.01 | 0 | -6.33 | -25.46 |
| | 0 | -0.04 | 3.58 | | 0 | 0.01 | 1 | 0 | 0 | 3.73 |
| 7 | 12.09 | 74.07 | -312.78 | | -1 | 0 | 0 | -12.09 | -74.06 | 312.76 |
| | -0.01 | 6.17 | -25.49 | | 0 | -1 | 0 | 0 | -6.22 | 25.68 |
| | 0 | 0.02 | 3.73 | | 0 | 0 | 1 | 0 | 0 | 3.83 |
| 8 | 12.05 | 72.91 | 313.04 | | -1 | 0 | 0 | -12.05 | -72.9 | -313.03 |
| | 0 | 6.13 | 25.7 | | 0 | -1 | 0 | 0 | -6.15 | -25.79 |
| | 0 | -0.01 | 3.83 | | 0 | 0 | 1 | 0 | 0 | 3.89 |
| 9 | 12.02 | 72.18 | -313.2 | | -1 | 0 | 0 | -12.02 | -72.18 | 313.2 |
| | 0 | 6.09 | -25.81 | | 0 | -1 | 0 | 0 | -6.1 | 25.85 |
| | 0 | 0.01 | 3.89 | | 0 | 0 | 1 | 0 | 0 | 3.93 |
| 10 | 12.01 | 71.71 | 313.3 | | -1 | 0 | 0 | -12.01 | -71.71 | -313.3 |
| | 0 | 6.06 | 25.86 | | 0 | -1 | 0 | 0 | -6.07 | -25.88 |
| | 0 | -0.01 | 3.93 | | 0 | 0 | 1 | 0 | 0 | 3.95 |
| 11 | 12.01 | 71.41 | -313.37 | | -1 | 0 | 0 | -12.01 | -71.41 | 313.37 |
| | 0 | 6.04 | -25.89 | | 0 | -1 | 0 | 0 | -6.05 | 25.9 |
| | 0 | 0 | 3.95 | | 0 | 0 | 1 | 0 | 0 | 3.97 |
| 12 | 12 | 71.21 | 313.41 | | -1 | 0 | 0 | -12 | -71.21 | -313.41 |
| | 0 | 6.03 | 25.9 | | 0 | -1 | 0 | 0 | -6.03 | -25.91 |
| | 0 | 0 | 3.97 | | 0 | 0 | 1 | 0 | 0 | 3.98 |
| 13 | 12 | 71.08 | -313.44 | | | | | | | |
| | 0 | 6.02 | -25.91 | | | | | | | |
| | 0 | 0 | 3.98 | | | | | | | |

The QR successfully finds the zeros at λ = 12, 6 and 4.

Now let's apply the QR to the polynomial used in **Example LR1**.

| Example QR2 | | | | *(Note:* $A_k = Q_k * R_k$ and $A_{k+1} = R_k * Q_k)$ | | | | | | |
|---|---|---|---|---|---|---|---|---|---|---|
| Iteration (k) | The A Matrices at Start of Each Iteration | | | | The Q Matrices (Orthogonal) | | | | The R Matrices (Upper Triangular) | |
| 0 | 0 | 1 | 0 | | 0 | -1 | 0 | | -5 | 11.5 | -2.5 |
| | 0 | 0 | 1 | | 0 | 0 | -1 | | 0 | -1 | 0 |
| | -5 | 11.5 | -2.5 | | 1 | 0 | 0 | | 0 | 0 | -1 |
| 1 | -2.5 | 5 | -11.5 | | -0.93 | -0.37 | 0 | | 2.69 | -4.64 | 10.68 |
| | 0 | 0 | 1 | | 0 | 0 | 1 | | 0 | -1.86 | 4.27 |
| | -1 | 0 | 0 | | -0.37 | 0.93 | 0 | | 0 | 0 | 1 |
| 2 | -6.47 | 8.91 | -4.64 | | -0.97 | 0.24 | 0 | | 6.67 | -9.64 | 4.94 |
| | -1.59 | 3.97 | -1.86 | | -0.24 | -0.94 | -0.23 | | 0 | -1.77 | 0.62 |
| | -0.37 | 0.93 | 0 | | -0.06 | -0.22 | 0.97 | | 0 | 0 | 0.42 |
| 3 | -4.45 | 9.64 | 7.01 | | -1 | -0.09 | 0 | | 4.46 | -9.47 | -6.9 |
| | 0.39 | 1.54 | 1.01 | | 0.09 | -0.99 | 0.06 | | 0 | -2.37 | -1.58 |
| | -0.02 | -0.09 | 0.41 | | -0.01 | 0.06 | 1 | | 0 | 0 | 0.47 |
| 4 | -5.23 | 8.61 | -7.46 | | -1 | 0.04 | 0 | | 5.24 | -8.69 | 7.52 |
| | -0.2 | 2.26 | -1.73 | | -0.04 | -1 | -0.01 | | 0 | -1.93 | 1.44 |
| | 0 | 0.03 | 0.47 | | 0 | -0.01 | 1 | | 0 | 0 | 0.49 |
| 5 | -4.91 | 8.78 | 7.63 | | -1 | -0.01 | 0 | | 4.91 | -8.75 | -7.61 |
| | 0.07 | 1.92 | 1.46 | | 0.01 | -1 | 0 | | 0 | -2.04 | -1.57 |
| | 0 | -0.01 | 0.49 | | 0 | 0 | 1 | | 0 | 0 | 0.5 |
| 6 | -5.04 | 8.65 | -7.64 | | -1 | 0.1 | 0 | | 5.04 | -8.67 | 7.65 |
| | -0.03 | 2.04 | -1.58 | | -0.01 | -1 | 0 | | 0 | -1.99 | 1.53 |
| | 0 | 0 | 0.5 | | 0 | 0 | 1 | | 0 | 0 | 0.5 |
| 7 | -4.99 | 8.69 | 7.65 | | -1 | 0 | 0 | | 4.99 | -8.69 | -7.65 |
| | 0.01 | 1.99 | 1.53 | | 0 | -1 | 0 | | 0 | -2.01 | -1.55 |
| | 0 | 0 | 0.5 | | 0 | 0 | 1 | | 0 | 0 | 0.5 |
| 8 | -5.01 | 8.67 | -7.65 | | -1 | 0 | 0 | | 5.01 | -8.67 | 7.65 |
| | 0 | 2.01 | -1.55 | | 0 | -1 | 0 | | 0 | -2 | 1.55 |
| | 0 | 0 | 0.5 | | 0 | 0 | 1 | | 0 | 0 | 0.5 |
| 9 | -5 | 8.68 | 7.65 | | | | | | | | |
| | 0 | 2 | 1.55 | | | | | | | | |
| | 0 | 0 | 0.5 | | | | | | | | |

The QR successfully finds the zeros at $\lambda = 0.5$, 2 and $-5$.

## Complex Eigenvalues (or Zeros)

For polynomials with real coefficients, complex zeros come in conjugate pairs. For real matrices, complex eigenvalues also come in conjugate pairs. It takes four elements in a matrix of real numbers to yield one complex pair. For example,

$$A = \begin{bmatrix} a & b \\ c & d \end{bmatrix}.$$

The eigenvalues of A are the roots of

$$\begin{vmatrix} (a - \lambda) & b \\ c & (d - \lambda) \end{vmatrix} = 0.$$

Assume that a matrix has four real eigenvalues and two complex pairs. The LR and QR methods must iterate until the form of the matrix resembles the following.

$$\begin{bmatrix} x & x & x & x & x & x & x & x \\ & x & x & x & x & x & x & x \\ & & x & x & x & x & x & x \\ & & x & x & x & x & x & x \\ & & & & x & x & x & x \\ & & & & & x & x & x \\ & & & & & x & x & x \\ & & & & & & & x \end{bmatrix}.$$

As can be seen, complex eigenvalues show up as *bulges* on the diagonal. Detecting them is a big part of the LR and QR convergence criteria. The following example demonstrates this.

Example QR3: Apply the QR program to the following polynomial $x^3 + 1.96x^2 + 5.528x + 0.43$.

$$A = \begin{bmatrix} 0 & 1 & 0 \\ 0 & 0 & 1 \\ -0.43 & -5.528 & -1.96 \end{bmatrix}.$$

The following table shows the results.

| Iteration (k) | The A Matrices at Start of Each Iteration | | | The Q Matrices (Orthogonal) | | | The R Matrices (Upper Triangular) | | |
|---|---|---|---|---|---|---|---|---|---|
| | | | | Example QR3 | (Note: $A_k = L_k * R_k$ and $A_{k+1} = R_k * L_k$) | | | | |
| 0 | 0 | 1 | 0 | 0 | -1 | 0 | -0.43 | -5.53 | -1.96 |
| | 0 | 0 | 1 | 0 | 0 | -1 | 0 | -1 | 0 |
| | -0.43 | -5.53 | -1.96 | 1 | 0 | 0 | 0 | 0 | -1 |
| 1 | -1.96 | 0.43 | 5.53 | -0.89 | -0.45 | 0 | 2.2 | -0.38 | -4.92 |
| | 0 | 0 | 1 | 0 | 0 | 1 | 0 | -0.2 | -2.51 |
| | -1 | 0 | 0 | -0.45 | 0.89 | 0 | 0 | 0 | 1 |
| 2 | 0.28 | -5.39 | -0.38 | -0.22 | 0.98 | 0 | -1.26 | 3.54 | 0.26 |
| | 1.14 | -2.24 | -0.2 | -0.91 | -0.2 | 0.37 | 0 | -4.72 | -0.33 |
| | -0.45 | 0.89 | 0 | 0.36 | 0.08 | 0.93 | 0 | 0 | -0.07 |
| 3 | -2.83 | -1.93 | 1.55 | -0.56 | -0.83 | 0 | 5.03 | 1.87 | -2.57 |
| | 4.16 | 0.94 | -2.06 | 0.83 | -0.56 | 0.01 | 0 | 1.07 | -0.12 |
| | -0.03 | -0.01 | -0.07 | -0.01 | 0 | 1 | 0 | 0 | -0.08 |
| 4 | -1.28 | -5.22 | -2.56 | | | | | | |
| | 0.88 | -0.6 | -0.12 | | | | | | |
| | 0 | 0 | -0.08 | | | | | | |

At iteration 4, the last row shows $\lambda_1$ = -0.08. The *bulge* shows the other two $\lambda$'s. They are the roots of the following equation.

$$\begin{vmatrix} (-1.28-\lambda) & -5.22 \\ 0.88 & (-0.6-\lambda) \end{vmatrix} = 0.$$

The solutions of this equation are $\lambda_2 = -0.94 + j2.12$, $\lambda_3 = -0.94 - j2.12$.

## The QR Method Expanded to Include *Matrix Deflation*

Suppose the following A matrix is detected during the iterations.

$$\left[\begin{array}{c|c} E & F \\ \hline 0 \bullet\bullet\bullet 0 & a \end{array}\right] \qquad \textit{Condition 1}$$

Here, the last row reveals that an eigenvalue has been found. As discussed in chapter 3, when the determinant of this matrix is expanded by minors, the last row and column can be dropped. QR iterations can continue with matrix A = E. This is called *matrix deflation*.

Now suppose the following A matrix is detected during the iterations.

$$\left[\begin{array}{c|cc} G & H & \\ \hline 0 \bullet\bullet\bullet 0 & a & b \\ 0 \bullet\bullet\bullet 0 & c & d \end{array}\right] \qquad \textit{Condition 2}$$

Here the last two rows reveal two eigenvalues. After computing them, the last two rows and columns can be dropped. The QR continues with A = G.

## The QR Method Expanded to Include *Shifting*

Bronson[12] discusses a theorem that says that if A has an eigenvalue $\lambda_1$, then one of the eigenvalues of the matrix $\left[A-\sigma_1 I\right]$ will be $(\lambda_1-\sigma_1)$. This is called *shifting*. If $\sigma_1 = \lambda_1$, then the matrix $\left[A-\lambda_1 I\right]$ will have an eigenvalue of zero. When this matrix is factored, the last row of R is forced to have all zeros. The last row of the reverse multiplication will also be all zeros. To regain similarity, the *shift* of $\lambda_1$ must be added back in. This is done via the statement $A=\left[A+\lambda_1 I\right]$. Hence the eigenvalue $\lambda_1$ will be revealed in the last row of A. Let's illustrate this.

$$A = \begin{bmatrix} 6 & 2 & -2 \\ 2 & 6 & -2 \\ -2 & -2 & 10 \end{bmatrix}$$

- Its eigenvalues are $\lambda = 4, 6$, and $12$. Set $\mathsf{shift} = 12*I$.

- Perform the shift, $A = \left[A - 12*I\right]$:

$$A_{shifted} = \begin{bmatrix} -6 & 2 & -2 \\ 2 & -6 & -2 \\ -2 & -2 & -2 \end{bmatrix}.$$

- Factor the shifted A into Q and R:

$$Q = \begin{bmatrix} -0.9 & -0.12 & 8 \\ 0.3 & -0.86 & -6 \\ -0.3 & -0.49 & 0 \end{bmatrix} R = \begin{bmatrix} 6.63 & -3.02 & 1.8 \\ 0 & 5.91 & 2.95 \\ 0 & 0 & 0 \end{bmatrix}.$$

- Reverse multiply $A = R*Q$:

$$\begin{bmatrix} -7.45 & 0.89 & 71.16 \\ 0.89 & -6.55 & -35.45 \\ 0 & 0 & 0 \end{bmatrix}.$$

- To restore similarity, shift back $A = \left[A + 12*I\right]$:

$$\begin{bmatrix} 4.55 & 0.89 & 71.16 \\ 0.89 & 5.45 & -35.45 \\ 0 & 0 & 12 \end{bmatrix}.$$

- From the last row, $\lambda_1 = 12$.

- From the bulge on the diagonal, the remaining $\lambda$'s are the roots of

$$\begin{vmatrix} (4.55-\lambda) & 0.89 \\ 0.89 & (5.45-\lambda) \end{vmatrix} = 0 \text{ which are } \lambda_2 = 4 \text{ and } \lambda_3 = 6.$$

[12] Bronson, R., *Linear Algebra* (Cambridge, MA: Academic Press, 1995).

Of course, exact shifting is not feasible because we don't know beforehand where the $\lambda$'s are. But experience has led to the development of several successful *guessing methods for shifting*.

Francis proposed the following strategy. Eigenvalues are computed from the submatrix in the lower-right-hand corner of the A matrix.

$$\begin{bmatrix} a_{n-1,n-1} & a_{n-1,n} \\ a_{n,n-1} & a_{n,n} \end{bmatrix} \qquad \textit{Guess 1}$$

Pick the eigenvalue of this submatrix that is closest to the element $a_{n,n}$. Use this eigenvalue as the *shift* for each QR iteration. The *shift* changes on each iteration, and it will become closer to a real eigenvalue of A. If the next eigenvalue is complex, this still works because *shifting* is only an aid in the iterative process.

There is another benefit from *shifting*. The QR does not diverge like the LR, but it can get into a limit-cycle. Sometimes after several iterations, the QR matrices begin to repeat. When this happens, there will be no convergence. Francis discovered that this limit-cycle could be interrupted by changing the method of *shifting* during the iteration process. His method is complicated.

I use a simpler method that is remarkably effective. If the first guess doesn't yield convergence after an arbitrary 100 iterations, the submatrix used to compute the *shift* is changed.

$$\begin{bmatrix} a_{n-1,n-2} & a_{n-1,n} \\ a_{n,n-2} & a_{n,n} \end{bmatrix} \qquad \textit{Guess 2}$$

## The QR Method Expanded to Include the Hessenberg Matrix

There is one final step that has become a part of most QR strategies. A Hessenberg matrix has all zeros below the subdiagonal.

$$\begin{bmatrix} x & x & x & x & x & x \\ x & x & x & x & x & x \\ 0 & x & x & x & x & x \\ 0 & 0 & x & x & x & x \\ 0 & 0 & 0 & x & x & x \\ 0 & 0 & 0 & 0 & x & x \end{bmatrix}$$

Being nearly triangular, it is common to transform the initial A to Hessenberg form before beginning QR iterations. Scheid discusses this transformation, which maintains similarity.[13]

---

[13] Scheid, F., *Numerical Analysis* (New York: McGraw-Hill Book Co., 1988).

The following is a flowchart of the program with all the expansions.

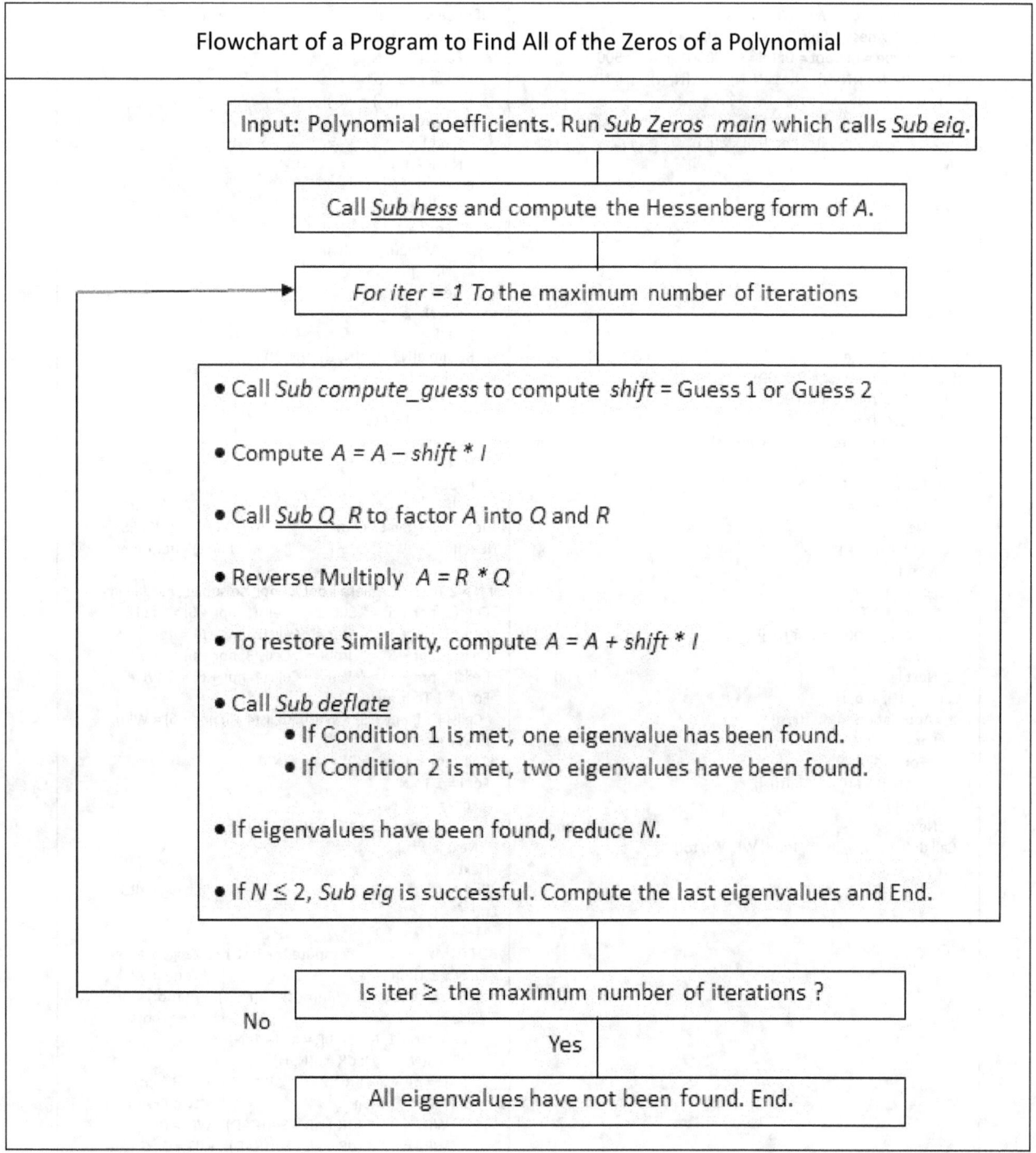

**Flowchart of a Program to Find All of the Zeros of a Polynomial**

Input: Polynomial coefficients. Run *Sub Zeros_main* which calls *Sub eig*.

Call *Sub hess* and compute the Hessenberg form of *A*.

*For iter = 1 To* the maximum number of iterations

- Call *Sub compute_guess* to compute *shift* = Guess 1 or Guess 2
- Compute $A = A - shift * I$
- Call *Sub Q_R* to factor *A* into *Q* and *R*
- Reverse Multiply $A = R * Q$
- To restore Similarity, compute $A = A + shift * I$
- Call *Sub deflate*
  - If Condition 1 is met, one eigenvalue has been found.
  - If Condition 2 is met, two eigenvalues have been found.
- If eigenvalues have been found, reduce *N*.
- If $N \leq 2$, *Sub eig* is successful. Compute the last eigenvalues and End.

Is iter $\geq$ the maximum number of iterations ?

No    Yes

All eigenvalues have not been found. End.

The program that implements these steps is listed in the following table. Chapter 5 explains how to use it.

| Subprogram eig Code, Page 1 of 6 | Subprogram eig Code, Page 2 of 6 |
|---|---|

```
Sub eig(npr, Ainput, WR, WI, tol, iter)
Dim q(), r(), Ahess(), Shift(), Asave(100, 100), a()
N = npr: itype = 0: iroot = 0: tol = 0.000001: niter = 500
ReDim q(N, N), r(N, N), Ahess(N, N), Shift(N, N), a(N, N)
For i = 1 To N
   For j = 1 To N
     a(i, j) = Ainput(i, j): Ahess(i, j) = a(i, j)
   Next j
Next i
If N < 3 Then GoTo BOTTOM
Call hess(N, Ahess)
For i = 1 To N
   For j = 1 To N
     a(i, j) = Ahess(i, j)
   Next j
Next i
For iter = 1 To niter                ''''' BEGIN QR ITERATIONS
  If iter = 100 Then tol = 0.00001
  If iter = 200 Then tol = 0.001
  If iter = 300 Then tol = 0.01
  Call compute_guess(a, N, iter, guess)          '' shift begin
    For i = 1 To N
      For j = 1 To N
       Shift(i, j) = 0
      Next j
       Shift(i, i) = guess
    Next i
    For i = 1 To N
      For j = 1 To N
        a(i, j) = a(i, j) - Shift(i, j)
      Next j
    Next i                              '' shift  end
  Call Q_R(N, a, q, r)
  a = Application.MMult(r, q)
    For i = 1 To N                       '''' unshift  begin
      For j = 1 To N
        a(i, j) = a(i, j) + Shift(i, j)
      Next j
    Next i                              '''' unshift  end
  Call deflate(N, a, itype, iroot, WR, WI, tol)
```

```
If itype <> 0 Then              '''''' Begin "A" Deflate
  If itype = 1 Then
   For i = 1 To N - 1
    For j = 1 To N - 1
      Asave(i, j) = a(i, j)
    Next j
   Next i
   N = N - 1
  Else                                '''  itype = 2
   For i = 1 To N - 2
    For j = 1 To N - 2
      Asave(i, j) = a(i, j)
    Next j
   Next i
   N = N - 2
  End If
  ReDim a(N, N), q(N, N), r(N, N)
  For i = 1 To N
    For j = 1 To N
      a(i, j) = Asave(i, j)
    Next j
  Next i
  itype = 0
End If                     '''''          End "A" Deflate
  If N <= 2 Then Exit For      '''''          SUCCESS
Next iter                      '''''       End QR Iterations

If N > 2 Then    ' here iroot <> npr, so sub eig has failed
  Cells(1, npr + 5) = " Sub eig": Cells(1, npr + 6) = " fails"
  Cells(2, npr + 5) = " iter =": Cells(2, npr + 6) = iter
  Cells(3, npr + 5) = " iroot =": Cells(3, npr + 6) = iroot
  Cells(4, npr + 5) = " WR =": Cells(4, npr + 6) = "  WI ="
  For i = 1 To iroot
    Cells(4 + i, npr + 5) = WR(i): Cells(4 + i, npr + 6) = WI(i)
  Next i
  Cells(npr + 4, npr + 2) = iter
  For i = 1 To N
    For j = 1 To N
      Cells(npr + 3 + i, j) = a(i, j)
    Next j
  Next i
  End                            '''' Sub eig FAILS
End If

BOTTOM:   '           compute the last 1 or 2 eigenvalues
  If N = 1 Then                      '''''   one REAL
   iroot = iroot + 1: WR(iroot) = a(N, N): WI(iroot) = 0
  Else                          ' N=2  two roots
   aR = a(N - 1, N - 1): bR = a(N - 1, N)
   cR = a(N, N - 1): dR = a(N, N)
   RAD = dR ^ 2 + aR ^ 2 - 2 * aR * dR + 4 * bR * cR
   If RAD >= 0 Then            '''''''   two REALs
     WR1 = 0.5 * (dR + aR + Sqr(RAD)): WI1 = 0
     WR2 = 0.5 * (dR + aR - Sqr(RAD)): WI2 = 0
   Else                        '''''''   one COMPLEX
     WR1 = 0.5 * (dR + aR): WI1 = 0.5 * Sqr(-RAD)
     WR2 = WR1: WI2 = -WI1
   End If
   iroot = iroot + 1: WR(iroot) = WR1: WI(iroot) = WI1
   iroot = iroot + 1: WR(iroot) = WR2: WI(iroot) = WI2
  End If                              '     N= 1 or 2
End Sub ' eig
```

| Subprogram eig Code, Page 3 of 6 | Subprogram eig Code, Page 4 of 6 |
|---|---|

```
Sub hess(N, AA)
 LA = N - 1: eps = 2 ^ (-52): T = 0
 For m = 2 To LA
  i = m: x = 0
  For j = m To N
   If Abs(AA(j, m - 1)) > Abs(x) Then
    x = AA(j, m - 1): i = j
   End If
  Next j
  If i <> m Then
   For j = (m - 1) To N
    y = AA(i, j)
    AA(i, j) = AA(m, j)
    AA(m, j) = y
   Next j
   For j = 1 To N
    y = AA(j, i)
    AA(j, i) = AA(j, m)
    AA(j, m) = y
   Next j
  End If
  If x <> 0 Then
   For i = (m + 1) To N
    y = AA(i, m - 1)
    If y <> 0 Then
     AA(i, m - 1) = y / x
     y = AA(i, m - 1)
     For j = m To N
       AA(i, j) = AA(i, j) - y * AA(m, j)
     Next j
     For j = 1 To N
       AA(j, m) = AA(j, m) + y * AA(j, i)
     Next j
    End If
   Next i
  End If
 Next m
 For jj = 1 To N - 2
  For ii = jj + 2 To N
    AA(ii, jj) = 0
  Next ii
 Next jj
End Sub ' hess
```

```
Sub compute_guess(a, N, iter, guess)
  If iter < 100 Then
     aR = a(N - 1, N - 1): bR = a(N - 1, N)
     cR = a(N, N - 1): dR = a(N, N)
  ElseIf iter >= 100 And iter < 200 Then
     aR = a(N - 1, N - 2): bR = a(N - 1, N - 1)
     cR = a(N, N - 2): dR = a(N, N - 1)
  ElseIf iter >= 200 And iter < 300 Then
     aR = a(N - 1, N - 1): bR = a(N - 1, N)
     cR = a(N, N - 1): dR = a(N, N)
  Else
     aR = a(N - 1, N - 2): bR = a(N - 1, N - 1)
     cR = a(N, N - 2): dR = a(N, N - 1)
  End If
    RAD = dR ^ 2 + aR ^ 2 - 2 * aR * dR + 4 * bR * cR
    If RAD >= 0 Then                      ''''' two REALs
       WR1 = 0.5 * (dR + aR + Sqr(RAD)): d1 = Abs(WR1 - a(N, N))
       WR2 = 0.5 * (dR + aR - Sqr(RAD)): d2 = Abs(WR2 - a(N, N))
       If d1 < d2 Then
         guess = WR1
       Else
         guess = WR2
       End If
    Else                                  ''''' one COMPLEX
       WR1 = 0.5 * (dR + aR): WI1 = 0.5 * Sqr(-RAD)
       guess = Sqr(WR1 ^ 2 + WI1 ^ 2)
    End If
End Sub ' compute_guess

Sub deflate(N, a, itype, iroot, WR, WI, tol)
 ctr = 0
 For j = 1 To N - 1
  If j < N - 1 Then
    If Abs(a(N - 1, j)) < tol Then ctr = ctr + 1  ' on each pass j < N - 1
    If Abs(a(N, j)) < tol Then GoTo NEXTJ
    Exit For                         ' return to QR  W/O success
  End If
                                        ''''' here j = N - 1
  If Abs(a(N, j)) < tol Then          ' begin j = N - 1 checks
    iroot = iroot + 1: WR(iroot) = a(N, N): WI(iroot) = 0 '''''  one REAL
    itype = 1: Exit For               ' return to QR to deflate
  End If
  If ctr = N - 2 Then                    '''''  two roots
    aR = a(N - 1, N - 1): bR = a(N - 1, N)
    cR = a(N, N - 1): dR = a(N, N)
    RAD = dR ^ 2 + aR ^ 2 - 2 * aR * dR + 4 * bR * cR
    If RAD >= 0 Then                     '''''  two REALs
       WR1 = 0.5 * (dR + aR + Sqr(RAD)): WI1 = 0
       WR2 = 0.5 * (dR + aR - Sqr(RAD)): WI2 = 0
    Else                                 '''''  one COMPLEX
       WR1 = 0.5 * (dR + aR): WI1 = 0.5 * Sqr(-RAD)
       WR2 = WR1: WI2 = -WI1
    End If
    iroot = iroot + 1: WR(iroot) = WR1: WI(iroot) = WI1
    iroot = iroot + 1: WR(iroot) = WR2: WI(iroot) = WI2
    itype = 2: Exit For                   ' return to QR to deflate
  End If                                  ' ctr=N-2
  Exit For                          '''  return to QR W/O success
NEXTJ:
 Next j
End Sub '  deflate
```

| Subprogram eig Code, Page 5 of 6 | Subprogram eig Code, Page 6 of 6 |
|---|---|

```
Sub Q_R(N, a, QM, RM)
Dim D(100), aQ(100, 100), Ident(100, 100), vh(), vhT(), H()
ReDim vh(N, 1), vhT(1, N), H(N, N)
 For i = 1 To N
   For j = 1 To N
     QM(i, j) = 0: RM(i, j) = 0: aQ(i, j) = 0: Ident(i, j) = 0
   Next j
     QM(i, i) = 1: Ident(i, i) = 1
 Next i
 K = 0
 For L = 1 To N            '''' Begin Transform
  K = K + 1
  If K = N Then
    D(L) = a(K, L): Exit For
  End If
  sarg = 0
  For i = K To N
    sarg = sarg + a(i, L) ^ 2
  Next i
  s = Sqr(sarg)
  If s = 0 Then
    D(L) = 0: GoTo nextL
  End If
  T = a(K, L): r = 1 / Sqr(s * (s + Abs(T)))
  If T < 0 Then s = -s
  D(L) = -s: a(K, K) = r * (T + s)
  For i = K + 1 To N
    a(i, K) = r * a(i, L)
  Next i
  For j = L + 1 To N
    T = 0
    For i = K To N
      T = T + a(i, K) * a(i, j)
    Next i
    For i = K To N
      a(i, j) = a(i, j) - T * a(i, K)
    Next i
  Next j
nextL: Next L               '''' End Transform
```

```
For i = 1 To N
  RM(i, i) = D(i)
Next i
For i = 1 To N - 1
  For j = i + 1 To N
    RM(i, j) = a(i, j)
  Next j
Next i
For j = 1 To N
  For i = j To N
    aQ(i, j) = a(i, j)
  Next i
Next j
For j = 1 To N - 1
  For i = 1 To N
    vh(i, 1) = aQ(i, j): vhT(1, i) = vh(i, 1)
  Next i
  vhvhT = Application.MMult(vh, vhT)
  For ii = 1 To N
    For jj = 1 To N
      H(ii, jj) = Ident(ii, jj) - vhvhT(ii, jj)
    Next jj
  Next ii
  Qtot = Application.MMult(QM, H)
  For ii = 1 To N
    For jj = 1 To N
      QM(ii, jj) = Qtot(ii, jj)
    Next jj
  Next ii
 Next j
End Sub  ' QR
```

## Appendix C:  A Program That Implements the LR Algorithm (See Appendix B)

```
Option Base 1
Sub LR_main()
Dim a(), L(), r()
N = Cells(1, 1)
ReDim a(N, N), L(N, N), r(N, N)              • Read in N and A(I, j).
For i = 1 To N
    For j = 1 To N
        a(i, j) = Cells(i + 1, j)
    Next j
Next i
For iter = 0 To 30                           • For iter = 0 To 13

    Cells(iter * (N + 1) + 1 + 1, N + 1) = iter
        For i = 1 To N
        For j = 1 To N
        Cells(iter * (N + 1) + 1 + i, j) = Application.Round(a(i, j), 2)
        Next j
    Next i
    Call L_R(N, a, L, r)                     • Call sub L_R(N, A, L, R)
    LR = Application.MMult(L, r)
    For i = 1 To N
        For j = 1 To N
        Cells(iter * (N + 1) + 1 + i, j + 2 * N + 2) = Application.Round(L(i, j), 2)
        Cells(iter * (N + 1) + 1 + i, j + 3 * N + 2) = Application.Round(r(i, j), 2)
        Cells(iter * (N + 1) + 1 + i, j + 4 * N + 2) = Application.Round(LR(i, j), 2)
        Next j                               • A ← R * L
    Next i
    a = Application.MMult(r, L)              • Next iter

Next iter
End Sub
```

```
Sub L_R(N, a, L, r)
    For i = 1 To N
    For j = 1 To N
    L(i, j) = 0: r(i, j) = 0                 • Subprogram L_R factors A into:
    Next j
    L(i, i) = 1                                 • L, a lower triangular matrix
Next i
For K = 1 To N - 1                              • R, an upper triangular matrix
    If a(K, K) >= 0 And a(K, K) < 0.000001 Then a(K, K) = 0.001
    If a(K, K) < 0 And a(K, K) > -0.000001 Then a(K, K) = -0.001    • See the following reference for LR factorization.
    For i = K + 1 To N                       Mathews, J., and K. Fink. Numerical Methods Using
        L(i, K) = a(i, K) / a(K, K)          MATLAB (Upper Saddle River, NJ: Prentice-Hall, 1999).
        For j = K + 1 To N
        a(i, j) = a(i, j) - L(i, K) * a(K, j)   • Note the poor man's work-around for division by zero.
        Next j
    Next i
Next K                                       • End Sub
For i = 1 To N
    For j = i To N
    r(i, j) = a(i, j)
    Next j
    Next i
End Sub
```

# Appendix D: A Program That Implements the QR Algorithm (See Appendix B)

```
Sub QR_main
    • Read in N and A(I, j)
    • For iter = 0 To (Niter = 11)
        • call Sub Q_R(N, A, Q, R)
        • A ← R * Q
    • Next iter
End Sub 'QR_main



Sub Q_R factors A into:

    • Q, a Householder Orthogonal matrix

    • R, an upper triangular matrix

    • See the following reference for QR factorization:
      Hager, W., Applied Numerical Linear Algebra
      (Englewood Cliffs, NJ: Prentice-Hall, 1988).

End Sub ' Q_R
```

```
Option Base 1
Sub QR_main()
Dim A(), Q(), R()
N = Cells(1, 1)
ReDim A(N, N), Q(N, N), R(N, N)
For i = 1 To N
    For j = 1 To N
        A(i, j) = Cells(i + 1, j)
    Next j
Next i
For iter = 0 To 11
    Cells(iter * (N + 1) + 1 + 1, 3 * N + 3) = iter
For i = 1 To N
    For j = 1 To N
        Cells(iter * (N + 1) + 1 + i, j) = Application.Round(A(i, j), 2)
    Next j
Next i
Call Q_R(N, A, Q, R)
A = Application.MMult(R, Q)
Next iter
End Sub ' QR_main
```

```
Sub Q_R(N, A, QM, RM)
Dim d(20), AQ(20, 20), Ident(20, 20), vh(), vhT(), H()
ReDim vh(N, 1), vhT(1, N), H(N, N)
 For i = 1 To N
   For j = 1 To N
     QM(i, j) = 0: RM(i, j) = 0: AQ(i, j) = 0: Ident(i, j) = 0
   Next j
     QM(i, i) = 1: Ident(i, i) = 1
 Next i
 k = 0
 For L = 1 To N              """ Begin Transform
   k = k + 1
   If k = N Then
     d(L) = A(k, L): Exit For
   End If
   sarg = 0
   For i = k To N
     sarg = sarg + A(i, L) ^ 2
   Next i
   s = Sqr(sarg)
   If s = 0 Then
     d(L) = 0: GoTo nextL
   End If
   T = A(k, L): R = 1 / Sqr(s * (s + Abs(T)))
   If T < 0 Then s = -s
   d(L) = -s: A(k, k) = R * (T + s)
   For i = k + 1 To N
     A(i, k) = R * A(i, L)
   Next i
   For j = L + 1 To N
     T = 0
     For i = k To N
       T = T + A(i, k) * A(i, j)
     Next i
     For i = k To N
       A(i, j) = A(i, j) - T * A(i, k)
     Next i
   Next j
nextL: Next L                """ End Transform
```

```
For i = 1 To N
    RM(i, i) = d(i)
Next i
For i = 1 To N - 1
    For j = i + 1 To N
        RM(i, j) = A(i, j)
    Next j
Next i
For j = 1 To N
    For i = j To N
        AQ(i, j) = A(i, j)
    Next i
Next j
For j = 1 To N - 1
    For i = 1 To N
        vh(i, 1) = AQ(i, j): vhT(1, i) = vh(i, 1)
    Next i
    vhvhT = Application.MMult(vh, vhT)
    For ii = 1 To N
        For jj = 1 To N
            H(ii, jj) = Ident(ii, jj) - vhvhT(ii, jj)
        Next jj
    Next ii
    Qtot = Application.MMult(QM, H)
    For ii = 1 To N
        For jj = 1 To N
            QM(ii, jj) = Qtot(ii, jj)
        Next jj
    Next ii
Next j
End Sub ' Q_R
```

# Appendix E: The Coordinate Transformation to Rotate the Conic Equation

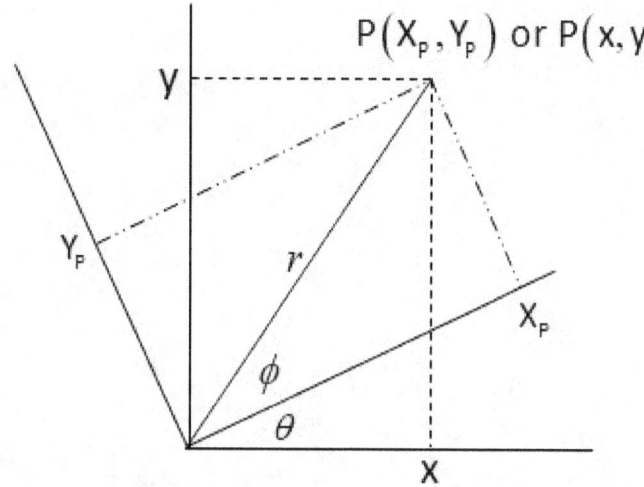

The equations in polar form:  (1) $X_p = r\cos\phi$

(2) $Y_p = r\sin\phi$

(3) $x = r\cos(\phi+\theta) = r\cos\phi\cos\theta - r\sin\phi\sin\theta$

(4) $y = r\sin(\phi+\theta) = r\sin\phi\cos\theta + r\cos\phi\sin\theta$

The equations in rectangular form:

- Substitute (1) and (2) into (3) and get → (5) $\boxed{x = X_p\cos\theta - Y_p\sin\theta}$
- Substitute (1) and (2) into (4) and get → (6) $\boxed{y = Y_p\cos\theta + X_p\sin\theta}$

Equations (5) and (6) are used to change the x-y coordinates into the $X_P$ - $Y_P$ coordinates. To invert equations (5) and (6),

- Rearrange (5) and get →  (7) $X_p\cos\theta = x + Y_p\sin\theta$

- Rearrange (6) and get →  (8) $Y_p = \dfrac{(y - X_p\sin\theta)}{\cos\theta}$

- Substitute (8) into (7) and get →  $X_p\cos\theta = x + \sin\theta\dfrac{(y - X_p\sin\theta)}{\cos\theta}$

- Rearrange the last equation and get →  $X_p\cos^2\theta = x\cos\theta + y\sin\theta - X_p\sin^2\theta$

- Rearrange the last equation and get →  (9) $\boxed{X_p = x\cos\theta + y\sin\theta}$

- Substitute (9) into (8) and get →  $Y_p = \dfrac{y - (x\cos\theta + y\sin\theta)\sin\theta}{\cos\theta}$

- Rearrange the last equation and get →  (10) $\boxed{Y_p = -x\sin\theta + y\cos\theta}$

Equations (9) and (10) are used to change the $X_P$ - $Y_P$ coordinates into the x-y coordinates. Equations (9) and (10) together form **Equation rotate**.